MICROORGANISMS IN FOODS:
their significance and methods of enumeration

Sponsored by

the International Committee on

Microbiological Specifications for Foods,

a standing committee of the International Association

of Microbiological Societies

Compiled and edited by

F. S. Thatcher, Chairman, ICMSF

D. S. Clark, Secretary-Treasurer, ICMSF

Contributors

M. T. Bartram, H. E. Bauman

R. Buttiaux, D. S. Clark

C. Cominazzini, C. E. Dolman

R. P. Elliott, M. M. Galton

H. E. Goresline, Betty C. Hobbs

A. Hurst, H. Iida

M. Ingram, K. H. Lewis

H. Lundbeck, G. Mocquot

D. A. A. Mossel, N. P. Nefedjeva

F. Quevedo, B. Simonsen

G. G. Slocum, F. S. Thatcher

University of Toronto Press

MICRO-ORGANISMS IN FOODS:
their significance and methods of enumeration

Reprinted 1973
Reprinted in 2018

Printed in Canada

ISBN 0-8020-1538-7
ISBN 978-1-4875-7270-9 (paper)
LC 73-2628

PREFACE

More and more countries are appreciating the need to assess the safety and quality of foods, because of wider recognition of the role of foods in spreading disease. This development has led to a rapidly growing diversity of microbiological methods for examining foods. These are applied mainly in official laboratories responsible for the safety or for the standards of quality of foods, but progressive food-processing firms often use the same or similar methods. A cause for concern is the fact that many food-processing establishments still sell products without knowledge of their microbial content.

Surveillance has become more necessary with the increase in international trade in foods and the hazards which could stem from the introduction of new techniques for mass production, rapid and widespread distribution, and introduction into commerce of foods from areas with endemic enteric diseases.[1] For example, the international nature of the salmonellosis problem is well documented.[2]

Manufacturers and control agencies need authoritative guidance on two problems long enveloped in uncertainty: (1) the significance of particular species or groups of microorganisms, when found in foods, and (2) microbiological specifications or standards. Statements related to either problem would have little meaning unless based on the use of effective methods for detection and enumeration of the pertinent microorganisms. Various microbiological specifications for foods, and a

1/See F. S. Thatcher, "The microbiology of specific frozen foods in relation to public health: Report of an international committee," J. Appl. Bacteriol. 26 (1963), 266.
2/See K. W. Newell, "The investigation and control of salmonellosis," Bull. World Health Org. 21 (1959), 279; L. W. Slanetz, C. O. Chichester, A. R. Gaufin, and Z. J. Ordal (eds.), "Microbiological quality of foods," Proc. Conf., Franconia, N.H., 27–29 Aug. 1962 (New York: Academic Press, 1963); E. J. Bowmer, "The challenge of salmonellosis: Major public health problem," Am. J. Med. Sci. 274 (1964), 467; U.S. Department of Health, Education, and Welfare, Proc. Natl. Conf. on Salmonellosis, 1964, Atlanta, Georgia (U.S. Public Health Service Publ. No. 1262, 1965); E. van Oye (ed.), The world problem of salmonellosis (Monographiae Biologicae, Vol. 13 [The Hague, the Netherlands: Dr. W. Junk Publishers, 1964]); B. C. Hobbs, "Contamination of meat supplies," Mon. Bull. Minist. Health Lab. Serv. 24 (1965), 123, 145. E. Vernon, "Food poisoning in England and Wales," Mon. Bull. Minist. Health and Public Health Lab. Serv. 25 (1965), 194.

plethora of methods, have been proposed. It is clearly desirable to try to establish internationally acceptable microbiological criteria and to reach agreement on the essential supporting methods.

The International Committee on Microbiological Specifications for Foods (ICMSF), a standing committee of the International Association of Microbiological Societies (IAMS), was formed in 1962 for this and related purposes. The "Agreed Programme" of the Committee is attached as Appendix I. The Committee seeks to aid in providing comparable standards of judgment among countries with sophisticated facilities for food control, and to offer useful procedures to the developing countries; to foster safe international movement of foods; and to dissipate difficulties caused by disparate standards and opinions about the significance of their microbial content. Fulfilment of such objectives should be of value to food industries, to the expansion of international trade in foods, to national control agencies, to the international agencies -more concerned with the humanitarian aspects of food distribution, and, eventually, to the health of the consuming public.

The Committee membership includes those associated with government departments of health and agriculture, with industry, and with universities, and having combined interests in research, official regulatory control, industrial control and product development, and education (see Appendix II, Membership). Nineteen different laboratories from 11 countries are represented.

The present document reports part of the outcome of deliberations by the ICMSF at meetings held in 1965 at Cambridge, England, and Vienna, Austria, in 1966 at Moscow, Russia, and in 1967 at London, England. Its purpose is threefold: (1) to describe the occurrence and significance of particular organisms in foods; (2) to offer recommendations on methods for the detection and enumeration of specific groups of food-borne microorganisms and (where feasible) for the detection of toxins causing food-poisoning; and (3) to describe the composition and methods of preparation of the recommended media, and to specify the related diagnostic reagents and tests.

The Committee recognizes two types of methods for different purposes: (a) precise methods of high sensitivity and reproducibility, to provide data for use in legal judgments or in arbitration related to public health or trade regulations; and (b) simpler, cheaper methods for routine surveillance, screening, and control. The Committee eventually intends to make recommendations for both types of methods, but this report relates only to the first, precise, type.

The intent has been to avoid proposing another set of "official" methods.

Instead, methods judged effective in different parts of the world have been assembled and their relative merits appraised in the light of the total experience of the Committee members.

The Committee wishes to emphasize that its proposals must be regarded as interim recommendations, and that interlaboratory testing is desirable.

A comprehensive literature review would be an undertaking beyond the scope of this presentation. Only salient references will be cited, with emphasis towards those of a review nature.

The terms used to express particular connotations for specific numbers of microorganisms in foods are adapted from definitions by the Food Protection Committee.[3] These definitions follow:

1 *A microbiological specification* is the maximum acceptable number of microorganisms or of specific types of microorganisms, as determined by prescribed methods, in a food being purchased by a firm or agency for its own use.

2 A *recommended microbiological limit* is the suggested maximum acceptable number of microorganisms or of specific types of microorganisms, as determined by prescribed methods, in a food.

3 A *microbiological standard* is that part of a law or administrative regulation designating in a food the maximum acceptable number of microorganisms or of specific types of microorganisms, as determined by prescribed methods.

The methods reported in this document have been adopted, with minor amendments to relate to irradiated foods, by the Expert Panel on Microbiological Specifications and Testing Methods for Irradiated Foods, which met in Vienna, Austria, in June 1965 and November 1967, under the joint sponsorship of the Food and Agriculture Organization and the International Atomic Energy Agency, and which included several members of ICMSF.

ACKNOWLEDGMENTS

The Committee expresses gratitude to the various companies within the food industry, to the World Health Organization, to the Pasteur Institute, and to the Carlo Erba Institute for Therapeutic Research, whose financial contributions (see Appendix III) have made possible the continuation of the work of the Committee, of which the present text is but one result; also to the respective departments of several national governments whose wisdom in supporting travel and ancillary costs for several members of the Committee we believe has been well rewarded.

3/See Food Protection Committee, U.S. National Academy of Sciences–National Research Council, *An evaluation of public health hazards from microbiological contamination of foods: A Report* (N.A.S.–N.R.C., Washington, D.C., Publ. 1195, 1964).

The Editors wish to thank all of the many scientists who contributed to the substance of this book and, in particular, Professor C. E. Dolman and Dr. M. Ingram for re-editing the text of Part I; and also the staff of the National Science Library of Canada for confirming the accuracy of the reference citations.

January, 1968   F.S.T.
                                                                                                     D.S.C.

ABBREVIATIONS OF
SOCIETIES AND AGENCIES

ACS   American Chemical Society

AFDOUS Association of Food and Drug Officials of the United States

AOAC   Association of Official Analytical Chemists

APHA   American Public Health Association

FAO   Food and Agricultural Organization of the United Nations

IAEA   International Atomic Energy Agency

IAMS   International Association of Microbiological Societies

ICMSF   International Committee on Microbiological Specifications for
        Foods

USFDA   United States Food and Drug Administration

WHO   World Health Organization of the United Nations

CONTENTS

PART I

PATHOGENIC
ORGANISMS
IN FOODS

COMMON
FOOD-POISONING
BACTERIA

Food may serve as a vehicle for the distribution of two major groups of organisms pathogenic to man: (1) those associated with endogenous animal infections transmissible to man (zoonoses) including bacterial, viral, fungal, helminthic, and protozoan species; (2) species which are exogenous contaminants of food, but which may cause infections or intoxications of man. The latter group includes the so-called food-poisoning organisms. Dolman (1957) has reviewed a number of criticisms of the term "food-poisoning," since it comprehends infective as well as entero-toxic processes, but it is retained here, chiefly because of its common usage. It implies a disease resulting directly from the consumption of food and after a relatively short incubation period. Food-poisoning of bacterial origin usually requires some proliferation of an infective pathogen or the elaboration of toxins within the food. Initial symptoms usually indicate a gastro-intestinal disturbance, but may change with the course of the disease (as in botulism).

The present work is confined to species of the second group (exogenous contaminants). The salmonellae, notable members of both groups, are included, but other important food-borne zoonotic pathogens, such as *Brucella, Pasteurella, Leptospira, Listeria, Erysipelothrix, Mycobacterium tuberculosis*, and *Bacillus anthracis*, which lack this dual role, will not be considered. The frequently prolonged incubation period of these pathogens often poses great difficulty in tracing the specific food source, and for many of them food would not be the major vehicle for infection of man inasmuch as most food-preservation treatments would destroy all but the sporing species, *B. anthracis*. Zoonoses are discussed in the report of the Joint WHO/FAO Expert Committee on Zoonoses (1959), and those acquired from meats, either by ingestion or contact, by Dolman (1957).

It is appropriate to distinguish between two types of food-poisoning bacteria: those, such as salmonellae and shigellae, which multiply in the intestinal tract and cause disease through infection of the host; and organisms such as *Staphylococcus* and *Clostridium botulinum* whose pre-formed toxins, present in food at the time of consumption, are the direct cause of illness. In addition, there is a third group of food-poisoning organisms for which the precise cause of illness is not yet known. An infectious role for *Clostridium perfringens* is not indicated (Hauschild *et al.*, 1967c), but no specific toxic substance has yet been established as the cause of the characteristic illness. Similar uncertainty prevails for *Bacillus cereus* and the enterococci. This last group, however, has an important property in common with the toxinogenic organisms: the ingested foods must, at some time, have allowed proliferation of the pathogen to very large numbers.

The differences in pathogenic action (infection or toxinogenesis) have influenced considerably the development of methods for the detection and enumeration of the respective groups of organisms. In general, sensitivity has been a major objective in devising methods for infective pathogens in foods, the inference being that disease may result from small numbers of organisms depending upon their specific virulence or upon the opportunity for proliferation. Further, the infective pathogens may involve very many patients as a result of secondary contamination and the development of asymptomatic carriers (see Salmonellae, below). Storage under conditions which prevent multiplication within the food will not confer safety against infective pathogens, but contributes greatly to protection against food-borne toxins.

The toxinogenic organisms have been of greatest concern usually: (i) where their toxins can be detected; (ii) when the organisms are present in relatively large numbers, and hence with an implied risk of the presence of toxins; and (iii) when present in foods of a nature likely to favour the production of toxins. In each case, a method sensitive enough to detect very sparsely distributed organisms has not been a primary objective. It is, nevertheless, recognized that modern developments in processing may require that more emphasis be placed on the detection of small numbers of potentially toxinogenic organisms in those foods from which competitive microflora have been removed, or which may be held under conditions that permit proliferation of toxinogenic species.

Accordingly, the food-poisoning bacteria to be considered here will be discussed in regard to their infective or toxinogenic characteristics, in so far as these are understood. Correspondingly, the methods to be described for the detection and enumeration of food-borne microorganisms have been selected largely in accord with the capacity of the method to meet the prior need, either sensitivity or indication of numbers, with sufficient accuracy to permit judgments within the purpose of the tests, whether conducted to assess potential hazard, degree of contamination, occurrence of proliferation, or compliance with a standard.

Infective food-poisoning bacteria

SALMONELLAE

Potential hazards

Salmonellosis is a gastro-intestinal infection usually of greatest severity in the very young or very old, though severe illness, with some mortality, may occur in any age group. The disease is characterized by an elevated temperature, diarrhoea, sometimes sufficient to lead to severe dehydration, intestinal pain, and perhaps vomiting. Consumption of contaminated foods is the most common cause. An incubation period of about 6 to 18 hours is usual. The duration of the disease may vary from a few days to a few weeks, depending in part on the efficacy of the therapy applied. A subsequent carrier state may persist in an asymptomatic patient for a few weeks to several months, and *Salmonella typhi* may be excreted intermittently throughout the carrier's lifetime.

The presence in foods of any serotype of *Salmonella* is potentially dangerous as a source of human disease, either directly upon consumption of food, or indirectly through secondary contamination of utensils, processing equipment, or other foods. A further risk arises through induction of the carrier state in food-handlers. Even though some serotypes, such as *S. pullorum* and *S. gallinarum*, may have relatively low virulence for man, nevertheless authentic human outbreaks from these pathogens of poultry have been recorded. In the present state of knowledge, all serotypes should be treated as potentially infective to man (Edwards and Galton, 1967; Edwards *et al.*, 1964). In addition, the group of allied organisms, arizonae, at one time considered as part of the genus *Salmonella*, should be suspect as potential pathogens and the present discussion on salmonellae should be regarded as applying also to this group (Edwards *et al.*, 1956).

"Cross-contamination" of foods within food-processing establishments is also a significant hazard, the inocula usually being derived from incoming untreated foods or from air-borne or surface-borne contact with powdered food ingredients, such as egg products. Mass-production technology and widespread distribution of foods can cause innumerable outbreaks of salmonellosis among extremely large populations (Thatcher, 1965). Exactingly thorough practices of factory sanitation are essential to control cross-contamination where foods or ingredients proved to contain salmonellae are being processed.

Control of salmonellosis

Effective control of salmonellosis requires action at several points in the prevailing epidemiological cycle (Newell, 1959; Bowmer, 1964; van Schothorst *et al.*, 1966; Edwards and Galton, 1967; Edwards *et al.*, 1964). This cycle involves the contamination of animal feed ingredients, such as fish-meal or protein concentrates, usually by inadequate sanitary control of the environment or of processing practices. The feeds represent one avenue for the contamination of cattle, hogs, and poultry, subsequently leading to contaminated meats, eggs (and chicks), egg products, and processed foods containing these basic ingredients. Farm environments, abattoirs, and food factories may become severely contaminated, leading to cross-contamination of other foods. Thus, milk may become contaminated with manure from infected cows, or with other components of the farm environment. Again, polluted water may lead to contamination of shellfish, farm environments, and domestic animals; infected food-handlers distribute salmonellae to many types of foods in factories, restaurants, and the home; offal and condemned carcasses reinfect the rendering plants that produce protein concentrates; and many other complex chains of transmission may be unravelled. To control the human disease, foods containing *Salmonella* must be prevented from reaching the consumer. The first essential step is to ensure that foods available to the public are removed from the market when they are found to contain salmonellae. Proof that such foods have caused salmonellosis should not be prerequisite to removal.

When feasible, foods known to be contaminated should be treated in a manner to destroy all salmonellae, before use as food either for man or animals. Food processing must be carried out in factories practising high standards of sanitation; contaminated ingredients should not be allowed to enter secondary food-processing factories; and cross-contamination must be prevented in the rendering and feed-mixing establishments.

Every effort should be made to prevent the movement of *Salmonella*-contaminated foods into food-processing factories, food-dispensing institutions, and homes. All such action and the supporting legislation are predicated upon the use of adequate analytical methods and sampling procedures to determine *Salmonella* at all points in the "cycle." This demonstrates emphatically the need for sensitive methods applicable to many substrates.

International surveillance of salmonellosis

A prerequisite to the control of salmonellosis is to know which food vehicles are primarily contributory to the disease, either in man or in domestic animals. Wide regional and national differences prevail in the efficiency of identifying salmonellae and of reporting the incidence of salmonellosis, the serotype involved, and the vehicle responsible. To aid in these aspects, the Committee recommends the development of an International Reporting Centre for Food-Borne Disease, initially having specific reference to salmonellosis. The proposed centre would have two main functions: (1) to collect, compile, and distribute information on the food vehicles, serotypes, and the related development of salmonellosis of man and animals; (2) to conduct training programmes to increase the level of competence in laboratory testing and epidemiological analysis.

An international reporting centre would not only aid development of effective control measures by national agencies, but would create an informed interest and awareness on the part of the food industry, thereby stimulating more effective testing and determination of the causative vehicle in outbreaks. Experience shows that once a food vehicle for *Salmonella* has been recognized, and the implications of this recognition are understood, with regard both to economics and to health, remedial measures are more likely to be implemented by the industries involved and by government.

In addition to the cost, a potential disadvantage in such a scheme was appreciated. Because of differences in the thoroughness of laboratory determinations and in the reporting of findings between nations, or between regional autonomies within a nation, the impression could be gained that contamination was most severe in those areas where testing was more effective. Such a bias would represent a penalty levied against active interest in reporting. Education to improve laboratory and epidemiological studies and to encourage accurate reporting was considered to be an essential remedial step.

As a working example of an integrated national reporting scheme, the epidemiological and associated reporting activities practised in the United Kingdom have been favourably cited. In the U.K., food-poisoning is a notifiable disease, and available information on all reported incidents is documented by the Public Health Laboratory Service and the Ministry of Health. The Epidemiological Research Laboratory, the *Salmonella* and Enteric Reference Laboratories, and various Working Parties of the Public Health Laboratory Service collaborate to produce annual reports giving details of causal agents in relation to persons, food vehicles, and places of infection (see Monthly Bulletin of the Ministry of Health and the Public Health Laboratory Service, 1950, and yearly thereafter).

The examination of foods and animal feeds for salmonellae is carried out on a large scale and, since 1962, annual data present human, animal, food, feed, and miscellaneous sources of the various serotypes of salmonellae (Vernon, 1965). These data are invaluable when searching for sources of *Salmonella* serotypes causing outbreaks of food-poisoning. In addition, numerous reports have been published on particular aspects of the salmonellosis problem, an activity in which contributors from many countries have, in recent years, developed a formidable body of literature. It is beyond the scope of this book to refer to more than a selection of reviews on the salmonellosis problem, such as those in the monograph *The World Problem of Salmonellosis* (van Oye, 1964) and those by Bowmer (1964), Newell (1959), the U.S. Department of Health, Education, and Welfare (1965), the Ministry of Agriculture, Fisheries and Food, the Veterinary Laboratory Service, and the Public Health Laboratory Service (1965), Hobbs (1965*a*), and Edwards and Galton (1967).

With a view to expanding the role of existing national activities in *Salmonella* investigation and reporting, the following group statement has been prepared. It has been the basis for exploratory contacts by the Committee with officers of who and of the Communicable Disease Center of the United States Public Health Service. The latter agency already has a valuable reporting scheme, in which a number of countries participate, and whose collective findings are published monthly (Salmonella Surveillance, 1964 *et seq.*).

AN INTERNATIONAL REPORTING CENTRE FOR SALMONELLOSIS

The Committee recognizes that salmonellosis is an international problem. Contaminated foods are the major vehicle for human salmonellosis. Since many food products are moving in international commerce, there is a great need for assembling data on an international basis.

The Committee recommends that an international organization (who) be approached regarding the establishment of an International Reporting System for salmonellae, and designation of a single Central International Reporting Centre, to be assisted by national and regional centres.

The Committee further recommends the development and establishment of national *Salmonella* reporting and surveillance centres in those countries where they do not already exist. One of the duties of these national centres would be to collect epidemiological information on *Salmonella* isolations from human and non-human sources, including foods, feeds, and animals. Where *Salmonella* typing centres already exist, collaboration with the reporting centre would be essential. The respective national centres would then report all isolates, together with other pertinent information regarding the sources of the cultures, to a designated International Centre, or to regional centres for transmission to the International Centre.

At the International Centre, the information received would be collated and reproduced in an International Salmonella Surveillance Report (monthly) for distribution to the reporting centres and other interested bodies.

To further the aims of this effort and to provide competence in identifying, reporting, and the surveillance of salmonellae, training facilities should be provided for personnel who will be engaged in this type of work. Such training should include particular attention to food and feed sources of human and animal infections. Training in the serological identification of salmonellae could be provided by the appropriate national Reference Laboratories.

The Committee further recommends that consideration be given to the development of similar international reporting systems for other important food-borne diseases.

General principles for the determination of salmonellae

It is not yet possible to recommend a single method for the isolation of salmonellae that would be fully satisfactory for all foods and for all sero-types. Prevailing methods are essentially similar in principle and embody the following series of procedural steps: pre-enrichment, enrichment, plating on selective agars, and taxonomic determination of colonies suspected of being *Salmonella*. Such complexity of procedure is chiefly due to the fact that salmonellae are mainly distributed, directly or indirectly, through faecal contamination. They are therefore usually present in foods in association with other organisms, particularly with members of the Enterobacteriaceae, such as *Escherichia, Enterobacter, Proteus, Citrobacter, Haffnia, Providencia,* and others. These genera, except for *Proteus,* comprise two groups often referred to as "coliforms" and "paracolons."

These groups of related enteric bacteria have many characteristics in common with *Salmonella* and are often present in much greater numbers than are the salmonellae. They may be better able to survive exposure to several adverse environmental factors such as heat, freezing, drying, and various antibacterial substances, and hence may be able to grow rapidly enough to inhibit development of salmonellae. On agar media they may prevent the formation of visible colonies, or mask the determinative characteristics of *Salmonella* colonies.

Present methods are designed to favour the multiplication of salmonellae while restricting the growth of competitive microorganisms. Non-selective enrichment is designed to favour the rapid recovery of salmonellae from a state of physiological inactivity or trauma, in which they are believed to exist in many foods, due to prolonged exposure to drying, freezing, high osmotic pressures and ionic strengths, or to low water activity, depending upon the type of food (Mossel, 1963). The media used for non-selective enrichment do not contain purposely added, selectively inhibitory chemicals. The selective enrichment media do contain such chemicals, the intent being to inhibit coliforms and other competitors to a greater extent than the salmonellae. Selective plating media also contain selective chemical inhibitors as well as indicators, the latter imparting characteristic colours either to the bacterial colonies or to the surrounding medium. Colonies with the appropriate characteristics for salmonellae, which are specific for a particular selective medium, are then subjected to tests for taxonomic verification and identification of serotype. The media employed in the taxonomic tests must be suitable for the reactions in question, but do not influence the sensitivity of a method for the isolation of salmonellae from foods.

Several variations exist in relation to non-selective enrichment, selective enrichment, and selective plating, which may be preferable for use with different types of foods. The Committee has embodied the principles of the four component steps, viz., non-selective enrichment, selective enrichment, selective plating, and taxonomic determination, in its proposed method for isolation of *Salmonella*, but recommends specific modifications and a choice of media depending on the nature of the food to be tested. The proposals represent a consensus based on much experience, despite a wide diversity of preferences that have not yet been fully resolved. The proposals are therefore somewhat arbitrary, but are offered because of the great importance of food-borne salmonellae in both domestic and international commerce, and because of the numerous requests received by members of the Committee for suitable methods for salmonellae. The Committee will undertake the necessary interlaboratory testing as a basis for future, possibly more definitive, recommendations.

In selecting a method for international use in relation to legal standards or for arbitration purposes, the need for sensitivity was generally conceded. "Short-cut" methods for routine control testing are valuable, but choice of such methods would probably be made nationally, by the industry, or by the official regulatory agency primarily concerned.

Regardless of the degree of refinement that *Salmonella* methodology might reach, the training, capability, attitude, and experience of the technician remain critical. It is also true that an insensitive method cannot give sound results even in the hands of an expert.

The fluorescent antibody procedure

Experience indicates that the fluorescent antibody technique for the determination of *Salmonella* in food shows promise (Haglund *et al.*, 1964; Georgala and Boothroyd, 1964; Silliker *et al.*, 1966). At the present stage of development, the method is best used as a presumptive test, particularly for the screening of large numbers of samples. Used in conjunction with cultural methods it can reduce the total effort in control analysis. But further research is required, particularly to improve the quality and range of fluorescent polyvalent antisera. The method requires the exercise of judgment by well-trained technicians and the best of optical equipment.

The large subjective element in the method demands wider experience in its application to different foods before specific recommendations can be offered for international adoption, though the method modified by adsorption of polyvalent O globulin with cultures of major interfering organisms (*E. coli and Citrobacter freundii*) was found to detect *Salmonella* in nine different foods without erroneous results (Insalata *et al.*, 1967).

Laboratory testing in food-processing plants

In at least one country, testing for *Salmonella* in food-processing premises is specifically prohibited by law. The basis for objection seemed to be the fear that maintaining cultures of a pathogen within a food plant could be dangerous. The Committee offers the following statement in support of its earlier plea for microbiological testing as an integral part of a food-processing system (Thatcher, 1963).

Since most if not all species of the genus *Salmonella* are potentially harmful to human health, and since many foods are known to contain these microorganisms, this Committee recommends that the manufacturers who process such foods should provide suitable laboratory facilities for *Salmonella* detection. Such a testing programme would be too large a task for existing governmental and private consultant laboratories to undertake.

The generally good public health record, in recent years, of the fluid milk industry in several technologically developed countries is largely due to that industry's acceptance of the need for bacteriological control of its products. It is believed that comparable benefits would follow the application of similar measures to other manufactured foods. Prompt performance of such tests permits the selection of raw materials and the release of finished products to be based on acceptable microbiological quality. Moreover, samples can be readily collected and management rapidly informed when the tests are conducted in a laboratory at or near the manufacturing plant.

The Committee is unaware of any episode in which contamination of foodstuffs has occurred from bacteriological cultures handled in a laboratory on the plant site; nevertheless, to guard against such remote contingencies, the proposed microbiological testing should be conducted in carefully designed laboratory premises, and carried out only by properly trained personnel under the supervision of a professionally qualified microbiologist. For companies too small to support the employment of a microbiologist, use of public analysts, perhaps supported through activities of industrial associations, is an alternative to be encouraged. Collaboration should be sought with inspection and control agencies of government. The advantages of laboratory control testing in food plants far outweigh any disadvantage or risk involved.

SHIGELLAE

The shigellae are not indigenous in foods and generally are not classified as a cause of outbreaks of food-borne infection of an explosive nature. They do, however, cause outbreaks of gastro-enteritis (shigellosis) and have been shown to be transmitted through food or water contaminated by human carriers (Keller and Robbins, 1956; Drachman *et al.*, 1960; Kaiser and Williams, 1962).

Shigellosis is caused by members of the genus *Shigella*, of which there are four serologically distinct groups, A–D. These groups refer, respectively, to the species *Sh. dysenteriae, Sh. flexneri, Sh. boydii, Sh. sonnei*. Each group contains several serotypes. *Sh. dysenteriae* usually causes the most severe illness, expressed by sudden onset of abdominal pain, tenesmus, pyrexia, and prostration. Diarrhoeal stools may quickly become composed mainly of blood and mucus. The illness associated with *Sh. sonnei* is less severe, only a small proportion of patients showing bloody stools; minor diarrhoea with recovery in from 48 to 72 hours is usual. The other groups tend to be intermediary in severity but show considerable variation.

Infection is normally by the oral route, often via the hands after contacting a contaminated object. Water and milk are the more common food vehicles, with the contamination source usually being active cases or carriers (Keller and Robbins, 1956; Drachman *et al.*, 1960; Kaiser and Williams, 1962). In some parts of the world food contamination via flies is an important factor in the prevailing shigellosis (Wilson and Miles, 1964). More recently, shigellae have been isolated from turkey droppings used as cattle feed, from hen feed and cracked eggs, from frozen and liquid egg melange, as well as from several environmental sites adjoining "henneries" (Cleare *et al.*, 1967). The extent to which processed foods may serve as vehicles is not clear, but since a small number of cells can cause infection, poor personal hygiene within the food-handling chain could be hazardous. Indeed Cruickshank (1965) points out that there is sufficient evidence to confirm that poor personal hygiene is a major factor in the spread of bacillary dysentery.

Studies on the survival of shigellae in foods have indicated that these organisms survived longest when holding temperatures were 25° C or lower. The length of survival varied with the food product. For example, *Sh. flexneri* and *Sh. sonnei* survived for over 170 days in flour and milk (Taylor and Nakamura, 1964), but survival was curtailed in acid products.

Very little work has been done to develop and evaluate enrichment and plating media specifically for the isolation of shigellae from foods. Procedures used for the recovery of shigellae from human faecal samples have been applied, such as Desoxycholate Citrate and Salmonella-Shigella agars (Lewis and Angelotti, 1964; Hartsell, 1951; Nakamura and Dawson, 1962). These are considered superior to MacConkey, Brilliant Green, or Bismuth Sulphite agars; selenite broths are preferable to tetrathionate (Wilson and Miles, 1964). Few enrichment media have been found to enhance significantly the growth of shigellae, although isolations from food have been made by use of Selenite F (Lewis and Angelotti, 1964) and more recently in some laboratories by use of Taylor's xylose lysine agars (Taylor, 1965). Obviously, research is needed to develop enrichment and plating media satisfactory for the isolation of shigellae from various food products.

THE ENTEROPATHOGENIC *ESCHERICHIA COLI* (EEC)

Several serologically identifiable strains of *E. coli* have been known for many years to cause infantile diarrhoea (Neter *et al.*, 1951; Taylor and Charter, 1952; Ewing *et al.*, 1957, 1963), but only recently have these EEC been recognized as food-borne pathogens. Costin *et al.* (1964) refer to 14 instances where EEC strains have caused illness in adults with epidemiological and clinical findings indicative of food-poisoning. The disease has been described as having a sudden onset after an incubation period of 7 to 12 hours with diarrhoea, abdominal pain, headache, and vomiting. Little or no fever was observed and the illness was of short duration, 24 hours or less. On the other hand, some outbreaks have been characterized by an incubation period of 3 to 4 days and more severe symptoms, including fever (39–40° C or 102–104° F) and prostration requiring hospitalization.

A variety of foods and several serotypes of EEC have been implicated: coffee substitute contaminated with serotype O86:B7:H34 (Costin *et al.*, 1964), Ohagi (red bean balls) with OB groups O55:B5 (Ueda *et al.*, 1959), stewed meat and gravy with O26:B6 (Koretskaia and Kovalevskaia, 1958), roast mutton with O26:B6 (Costin *et al.*, 1960), and pork and chicken with a number of serotypes (Smith *et al.*, 1965), among others. With some of these the evidence is quite clear, but with others not so convincing.

It is especially significant that infected adults become carriers for short periods after exposure and therefore are potentially dangerous to any infants with whom they may associate. Furthermore, recent work by Newell *et al.* (1966) and Smith *et al.* (1965) suggested that EEC might be introduced into a household by food purchased from markets. Recent work has also demonstrated the occasional occurrences of EEC in market foods, in the faeces of food-handlers (Hall and Hauser 1966; Hall *et al.*, 1967), and in pig faeces and carcasses (Filippone *et al.*, 1967). The detection of relatively large numbers of a particular EEC serotype in a food and in the faeces of consumer patients allows the laboratory to establish the aetiology and the vehicle involved.

In addition, the detection of even low numbers of EEC in food such as instant non-fat dry milk or formulated baby food may reveal a public health hazard as significant as the detection of salmonellae in such a food.

Methods specific for the enumeration of these strains from foods have received relatively little investigation. For specimens containing larger numbers of *E. coli*, direct plating is employed using the media recommended for *E. coli* (see Part II, p. 72), such as Eosin Methylene-Blue or MacConkey agars. Edwards and Ewing (1962, p. 70) additionally recommend the use of Blood Agar plates, a practice also employed by Filippone *et al.* (1967). For sparse populations, a tentative recommendation is to use a non-selective medium such as Peptone Water followed by a "mild" selective enrichment such as Lauryl Sulphate Tryptose Broth (Hall, Brown, and Lewis, 1967). Cultures giving an indication of *E. coli* at this stage are plated on a selective agar, such as Eosin Methylene-Blue or MacConkey agars, and a number of representative colonies are tested with O and OB antisera specific for the group. Further serotyping, where necessary, is carried out by well-standardized procedures (Edwards and Ewing, 1962).

More recently the fluorescent antibody (FA) technique (see Salmonellae, p. 10), used in conjunction with cultural tests, has been applied to the determination of EEC in faeces, organs, and carcasses of pigs (Filippone *et al.*, 1967). The FA technique was judged to be sensitive, but positive indications required confirmation.

VIBRIO PARAHAEMOLYTICUS

Vibrio parahaemolyticus is a Gram-negative halophilic organism well known in Japan as a cause of the food-poisoning syndrome associated with the consumption of raw fish and shellfish (Kawabata and Sakaguchi, 1963). The food-poisoning due to this organism is thought to be of an infectious type resembling salmonellosis. The enteropathogenicity of the organism has been established by several human volunteer tests, and positive results have been obtained only by administering living cultures. As yet no enterotoxin-like substance has been demonstrated in culture filtrates.

The disease occurs mainly during the warm summer months. The incubation period ranges from 6 to 24 hours (average 14–20 hours). The disease usually begins with a violent epigastric pain accompanied by nausea, vomiting, and diarrhoea. In severe cases, mucus and blood are observed in the faeces. Fever ranging up to 39° C (102° F) is observed in most cases. Because of these symptoms, the illness is often erroneously diagnosed as dysentery (Aiso and Matsuno, 1961).

Organisms causing these symptoms are usually called "pathogenic halophiles" and have variously been referred to as *Pasteurella parahaemolytica* (Fujino *et al.*, 1953), *Pseudomonas enteritis* (Takikawa, 1958), and *Aeromonas* (*Oceanomonas*) *parahaemolytica* (Miyamoto *et al.*, 1961). Eventually, they were recognized as members of a group belonging to the genus *Vibrio*, among which two closely related biotypes are recognized (Sakazaki *et al.*, 1963). Epidemiological and aetiological evidence suggests that only biotype 1 is enteropathogenic (Zen-Yoji *et al.*, 1965), and this is named *V. parahaemolyticus*. The second biotype, *V. alginolyticus*, may also cause food-poisoning in man under special circumstances.

These species have been frequently found in coastal sea-waters of Japan. Recently they were found in samples of sea-water, fish, and plankton collected at Honolulu, Hawaii, and at Hong Kong and Singapore. The water-borne organisms are believed to contaminate the living fish and shellfish. After the fish have been caught, the organisms proliferate rapidly, especially at the high temperatures during summer, when numbers sufficient to cause food-poisoning are attained within a few hours (Miyamoto *et al.*, 1961).

Recently two monographs on *V. parahaemolyticus* have been published, which contain a complete bibliography on this organism: *Vibrio parahaemolyticus*, Series I, 1963, edited by T. Fujino and H. Fukumi, published by Isseido Co., Tokyo, Japan (written in Japanese); Series II, 1967, edited by T. Fujino and H. Fukumi, published by Nayashoten Co., Tokyo (written in Japanese).

Enterotoxic food-borne pathogens

AEROBIC ENTEROTOXIC SPECIES: THE STAPHYLOCOCCI

Staphylococcal food-poisoning

A food-poisoning syndrome characterized by nausea, vomiting, diarrhoea, general malaise, and weakness, and in more severe cases with collapse and other signs of shock, which begins in from 30 minutes to 3 hours after ingestion of food, is usually due to consumption of a food that contains large numbers of staphylococci (Dolman, 1943; Dack, 1956).

The symptoms are caused by specific polypeptides, which function as emetic toxins, known as enterotoxins. These are freed into the food matrix by certain strains of *Staphylococcus* (Bergdoll *et al.*, 1959; Casman *et al.*, 1963; Casman and Bennett, 1965; Bergdoll, 1963; Joseph and Baird-Parker, 1965; Thatcher and Robinson, 1962; Thatcher, 1966).

The species of *Staphylococcus* known with certainty to contain many strains capable of causing food-poisoning is *Staph. aureus*. Almost all strains of this species produce the enzyme coagulase, but only a small proportion produce enterotoxin (Casman and Bennett, 1965). On the other hand, coagulase-negative strains have been found which produce enterotoxin (Dolman, 1942; Bergdoll *et al.*, 1967), and coagulase-positive staphylococci from patients receiving antibiotic treatment may not show this enzyme on first isolation. Different clones within a culture may vary in their capacity to produce pigment and also coagulase. Nevertheless, the coagulase test remains an essential part of the determination of *Staph. aureus*.

Four specific enterotoxins called enterotoxins A, B, C, and D are now serologically identifiable (Casman, 1966; Borja and Bergdoll, 1967; Casman *et al.*, 1967), but additional emetic toxins are under investigation. The toxin most commonly implicated in food-poisoning seems to be enterotoxin A, though a large outbreak from cheese was recently shown to be due to enterotoxin C (Casman, 1966). Anti-enterotoxic sera have been prepared against each of the four enterotoxins. Currently, immuno-diffusion (Casman and Bennett, 1965; Borja and Bergdoll, 1967) and haemagglutination-inhibition tests (Morse and Mah, 1967; Johnson, Hall, and Simon, 1967) are in experimental use for detection of the specific enterotoxins. However, the monovalent antisera are not yet available for general use, and for the present, the specific detection of enterotoxigenic strains rests upon use of human volunteers or of test animals such as monkeys or cats. All three of these experimental animal species present their peculiar problem (Dolman, 1943; Thatcher and Robinson, 1962; Bergdoll, 1963). Hence, when seeking to determine potential food-poisoning strains, until more suitable methods are available for the detection of the enterotoxins, it may be assumed that *in general*, unless a strain is coagulase-positive, it is unlikely to produce enterotoxin.

The determination of the phage pattern of staphylococci is a useful epidemiological aid. The strains more commonly implicated in food-poisoning are members of phage groups III and IV (Parker and Lapage, 1957), but the production of enterotoxin A among a large number of strains has been shown to bear little relationship to the phage pattern (Levy *et al.*, 1953; Casman and Bennett, 1965). Strains untypable by standard phages may also produce enterotoxins (Casman and Bennett, 1965; Thatcher and Erdman, 1966).

Processed foods in which a large population of staphylococci have been destroyed by heating may nevertheless cause food-poisoning owing to survival of the heat-resistant enterotoxins. Secondary contamination with other strains of staphylococci, after the heating process, may add to the difficulty in determining the causative strain. Again, the staphylococci which initially produced toxin in a food may become overgrown by other microorganisms (Dolman, 1943), so that analysis may not reveal a sufficient number of staphylococcal organisms to indicate a diagnosis of staphylococcal food-poisoning without further evidence. In certain foods, also, staphylococci may die rapidly and eventually toxin may persist in the presence of small numbers of viable staphylococci.

The diverse significance of staphylococci in foods

A judgment as to the relative significance of staphylococci in a particular food needs to be made with care. Many raw foods, including milk and unpasteurized dairy products, may contain very large numbers of staphylococci. Their significance in such foods relates chiefly to their capacity to produce enterotoxins if the foods are held at appropriate temperatures. In some foods, toxin formation is rare in relation to the frequency of the presence of staphylococci, even though large numbers of staphylococci may occur commonly (Meyer, 1953). This is particularly true for Cheddar type cheese made from unpasteurized milk (Thatcher *et al.*, 1956). Such foods can serve as a vehicle for widespread distribution of *Staphylococcus*, though the extent to which this route contributes to human infection has not been conclusively demonstrated. The possibility of an infective hazard should be entertained, however, since milk and dairy products, unless heat-treated, frequently contain large numbers of staphylococci of phage patterns identical with those known to have become established in human carriers and which have frequently caused severe infections (Thatcher *et al.*, 1959; Munch-Petersen, 1960; Munch-Petersen and Boundy, 1962).

In foods of a nature which will not allow proliferation of staphylococci the chief concern over the presence of staphylococci should be whether the foods are likely to be used as ingredients of other foods, such as mayonnaise, cream toppings or fillings of cakes and desserts, grated cheese on cooked dishes, wherein staphylococci might grow readily and produce enterotoxin.

Apart from the foregoing role of viable staphylococci in dried foods, there is a less frequent but clearly demonstrated hazard. Fluid foods, particularly milk, if held for a few hours at temperatures that allow growth of staphylococci, may permit elaboration of significant amounts of enterotoxin. Subsequent pasteurization and a drying process will kill most or all of the staphylococci, but the heat-resistant enterotoxins will remain. Outbreaks of staphylococcal food-poisoning from reconstituted powdered milk (Anderson and Stone, 1955; Armijo et al., 1957) and from a powdered malted-milk drink have been clearly established (Dickie and Thatcher, 1967), even though the dried products contained few or no viable staphylococci.

In cooked or processed foods staphylococci are good indicators of sanitation, particularly as regards control of the personal hygiene of factory workers. Factory personnel may contribute staphylococci from respiratory infections, suppurative lesions (boils, infected cuts, abrasions, etc.), from the nostrils of "carriers" (usually via the hands), or by coughing, sneezing, or expectoration, particularly while suffering from a throat or bronchial infection, often seen as a sequel to the common cold. Staphylococci tend to be resistant to drying and thus have some value in assessing the adequacy of surface disinfection procedures used in food factories (Thatcher, 1955). For products of particular factories, the staphylococcal content of precooked frozen dinners has been shown to be a good indicator of the quality of sanitation (Shelton et al., 1962).

ANAEROBIC ENTEROTOXIC SPECIES: *CLOSTRIDIUM BOTULINUM*

Food-poisoning by Clostridium botulinum

Food-borne botulism is an intoxication due to consumption of food containing botulinus toxin formed during previous growth of *C. botulinum*. It is characterized by a non-febrile, neuroparalytic syndrome and a generally high mortality rate. Disturbances of vision and difficulty in speaking and swallowing are among the more distinctive features, followed by progressive weakness and respiratory and cardiac failure in severe cases. Abdominal distension and pain, with or without vomiting, are commonly present, but constipation is more usual than diarrhoea.

The incubation period is seldom less than 6 hours, and the onset of symptoms may be delayed for 12 to 24 hours, or even longer. In the severest cases, particularly of type E botulism, death can occur in 20 to 24 hours after ingestion of the implicated food, but is more liable to occur after an illness of two or three days, or even longer than a week.

Clostridium botulinum is a strictly anaerobic species, which forms highly resistant spores. There are six known types, designated A to F,

which are distinguished by monovalent antitoxin-neutralization tests.
Types A, B, E, and F toxins are established as dangerous when ingested
by man, but most outbreaks of human botulism have been caused by
types A, B, and E.

Spores of this species are widely but irregularly scattered in the soil and
off-shore waters of many regions. Hence, for example, vegetables grown
in these areas, or fish feeding in such waters, are liable to contamination
by the spores. The heat resistance of these spores, while variable, is
generally high for types A and B strains, whereas type E spores tend to
be heat-susceptible. This factor greatly influences not only the epidemiology of the disease, and the kinds of foods involved, but also laboratory diagnostic techniques. All types can produce toxin in a variety of
foodstuffs, sometimes at relatively low temperatures, e.g. down to 10° C
for types A and B strains or 4° C for type E strains; and also at unexpec-
tedly low pH levels, such as the range 4.0 to 5.0 for some type E strains.

International symposia on botulism have been held at Cincinnati (Lewis
and Cassel, 1964) and Moscow (Ingram and Roberts, 1966). The
published proceedings of these conferences contain reports on most aspects
of the epidemiology, control, and laboratory diagnosis of human botulism.

Detection and control

Contemporary methods of detecting the organism, and identifying its
serotypes A to F, are based on the protection of test animals from the
specific heat-labile toxins of the serotypes, by use of homologous antisera.
Because the organism culturally resembles certain common non-toxinogenic clostridia (particularly *C. sporogenes*), special tests for the toxins
continue to be essential, despite extensive efforts to discover alternatives.

The choice of procedure will be influenced by the urgency of the test.
Where a particular food is suspected of causing botulism, it is essential
to complete the test with minimum delay, in part to warrant the drastic
action necessary to prevent further distribution or use of the suspect
source, in part to ensure use of the appropriate therapeutic antiserum. The
test may need minor modification according to the amount of specimen
available; the only specimen may be an empty container retaining only
traces of the food, or a small food residue recovered from household waste.

In urgency, a direct test for toxin is the objective. Usually, one resorts
to isolation and identification of the organism only when specimens are
inadequate for demonstration of toxin. However, isolation and identification of the organism becomes necessary: (1) when the mice under test
have died with atypical symptoms or after an unusually short or long
incubation period or (2) when all type-specific antitoxins fail to protect

the inoculated animals. The latter situation might imply the presence of either multiple toxins or a hitherto unrecognized type of toxin—from *C. botulinum* or of quite different origin.

To prevent the development of botulinus toxin is an essential objective in food-preserving processes whenever the food does not possess properties preventing growth of the organism—e.g., pH below 4.5, low water activity, or a sufficient concentration of curing salts. Methods of food preservation must either (i) kill all the spores, (ii) prevent their development, or (iii) provide an environment which prevents growth and toxin formation. For processed foods which will allow formation of toxin, two requirements are essential: (*a*) in a heat process, times and temperatures suitable to inactivate *C. botulinum* should be specified by competent experts, for each food formulation, container size, and method of processsing; (*b*) foods in which *C. botulinum* has not been inactivated should be kept frozen, or at temperatures below the limits for growth and toxin formation (with type E, this temperature is about 4° C).

Food-poisoning bacteria, specific cause uncertain

CLOSTRIDIUM PERFRINGENS

Certain strains of *Clostridium perfringens* cause a relatively mild malaise, characterized by abdominal pain and diarrhoea, the onset of the symptoms usually occurring within 8 to 18 hours after ingestion of the food responsible (McClung, 1945; Hobbs *et al.*, 1953; Dische and Elek, 1957; Nygren, 1962). The factor responsible for inducing the disease is not yet known. The organism is common in foods, in the soil, and in mammalian faeces. A population of about one million organisms per gram of food (Dische and Elek, 1957) is necessary before disease is likely to result from the ingestion of such food.

The criteria generally accepted for implicating *C. perfringens* in food-poisoning outbreaks are the clinical details, history of the outbreak, epidemiology, and the mode of preparation of the suspected food. An additional requirement by some workers in this field is that the strain isolated from the suspected food should match the serotype of the strain isolated from the patient's stools. Others are of the opinion that adequate indication of the causative organism may be derived from positive identification of *C. perfringens* in large numbers from suspected food.

Strains of *C. perfringens* type A might be grouped as follows: (i) "typical" food-poisoning strains, which are weakly haemolytic or non-haemolytic on Horse-Blood Agar and whose spores are markedly heat-resistant, having a D value[1] of the order ½ to 1 hour at 100° C; (ii) β-hae-molytic strains which produce heat-susceptible spores; (iii) weakly or non-haemolytic strains which produce heat-susceptible spores.

Strains of group (i) have been recognized as causing food-poisoning for many years (Hobbs, 1965b). More recently, a variety of heat-suscep-tible strains were implicated in food-poisoning outbreaks (Hall et al., 1963; Taylor and Goetzee, 1966; Sutton and Hobbs, 1968). Strains of group (ii) were shown to produce the food-poisoning syndrome in lambs which are now used as experimental animals (Hauschild et al., 1967c), and a strain of group (iii) was shown to produce experimental food-poisoning in humans (Hauschild and Thatcher, 1967). Some type A food-poisoning β-haemolytic strains (group ii) can also cause gas-grangrene (Hauschild and Thatcher, 1968). A relationship between the acid resistance of growing cells and the capacity to cause food-poisoning has been demonstrated among a number of strains by Hauschild et al. (1967b). The general consensus now is that differentiation between clas-sical and food-poisoning strains is no longer valid.

With few exceptions, the food vehicles for C. perfringens in the United Kingdom have been uncured meats or meat products (Hobbs, 1965b). The large majority of strains isolated were of group (i). The incriminated food vehicles in the United States have included a number of prepared or semi-prepared mixtures, including meat dishes and beans (Hall et al., 1963).

Three methods are available for the quantitative determination of C. perfringens. The method employed in the United Kingdom is based on the assumption that the strains with heat-resistant spores are more likely to be involved in food-poisoning than those which are easily killed by cooking, and seeks to identify the haemolytic character on the surface of Blood-Agar media (Hobbs et al., 1953; Hobbs, 1965b). A modification of this procedure applies tests for both spores and vegetative cells (Beerens and Tahon-Castel, 1965). Investigators in North America use the Sul-phite Polymyxin Sulphadiazine (SPS) agar of Angelotti et al. (1962), which is highly selective for clostridia. A preference is expressed in some laboratories for the exclusion of sulphadiazine from the plating medium, because of the frequent inhibition of spores of C. perfringens by the sul-phonamide (Mossel et al., 1965). However, of a large number of strains derived from American and European sources, none were inhibited by laboratory preparations of Angelotti's SPS agar (Hauschild et el., 1967a). The use of an appropriate diluent has been shown to be very important

1/The D value refers to the time, in minutes at a stated temperature, required to destroy 90 per cent of a population of spores.

for enumerative purposes, particularly when using commercial media. Peptone solution is the most satisfactory diluent (Hauschild *et al.*, 1967*a*).

BACILLUS CEREUS

Bacillus cereus has the potentiality to cause a mild form of food-poisoning with marked similarity in symptoms and time of onset to that caused by *Clostridium perfringens* (Christiansen *et al.*, 1951; Hauge, 1955; Clarenburg and Kampelmacher, 1957; Nikodemusz, 1958; Bodnar, 1962; Nygren, 1962). The causal substance has not been conclusively demonstrated. *B. cereus* is an aerobic spore-forming organism common in soil, on vegetation, and in many foods, raw or processed. Inadequate refrigeration of moist, cooked, protein foods is the essential factor in allowing proliferation of this organism, which causes illness only when ingested in very large numbers. The determination of relatively small numbers of *B. cereus* in a food, for example less than 10^6 per gram, is probably not sufficient evidence to warrant an assumption of food-poisoning by this organism.

THE ENTEROCOCCI

The enterococci, which with other species make up Lancefield's Group D of the genus *Streptococcus*, include the species *Streptococcus faecalis*, *Strep. faecium*, and their varieties (Sherman, 1937; Niven, 1963; Deibel, 1964). A classification scheme for these organisms is presented in Table 5, p. 90. Foods held at temperatures in the mesophilic range may come to contain very large numbers of these organisms, and occasionally such foods have been thought to be involved in food-poisoning. Buchbinder *et al.* (1948) described such cases, and Hartman *et al.* (1965) reviewed several similar indications.

The latter authors tend to concur with Niven (1963) and with Deibel (1964), however, that decisive proof of the role of enterococci as a cause of food-poisoning has not yet been established; nor is it clear which specific types may be involved. Synergistic association with some other organism may be necessary (Hartman *et al.*, 1965).

INDICATOR ORGANISMS

Methods used to determine pathogens are less effective when the pathogen is sparsely distributed in a food, particularly when the food is heavily contaminated with other organisms. Even when sensitive methods are available, the cost and time involved may be prohibitive. Such difficulties have led to the widespread use of groups (or species) which are more readily enumerated, and whose presence in foods (within numerical limits) is considered to indicate exposure to conditions that might introduce hazardous organisms and/or allow proliferation of pathogenic or toxinogenic species. The groups or species so used are called "indicator" organisms, and have value in assessing both the microbiological safety and quality of foods.

Several different indicator organisms have been used for different purposes. The basis for such use, and an assessment of their significance in various foods, will be considered briefly.

The objective in using bacteria as indicators of unsanitary practice is to reveal possible conditions of treatment which imply a potential hazard, not necessarily present in the particular sample examined, but liable to be present in parallel samples. Lewis and Angelotti (1964) have recently reviewed the methodology for food-borne indicator bacteria, to provide basic information to help ICMSF in compiling this present book.

Aerobic mesophilic bacteria

Most foods (except, for example, fermented products) are regarded as unwholesome when they have a large population of microorganisms, even if the organisms are not known to be pathogens and they have not altered the character of the food noticeably. For this there are several reasons.

(i) High viable counts often indicate contaminated raw materials, unsatisfactory sanitation, or unsuitable time/temperature conditions, during production or storage, or a combination of these. If the organisms are mesophiles (i.e., grow best at temperatures near 37° C), this means that conditions have existed which might favour the occurrence of pathogens in large numbers.

(ii) Various common bacteria not usually regarded as pathogens (e.g., faecal streptococci, *Proteus*, pseudomonads) have been suspected of causing poisoning when excessive numbers of living cells were present in food. It therefore seems wise to prevent the development of a high viable count, until such occurrences are better understood. Because the pathogenic and suspect species are almost all mesophilic bacteria, attention is concentrated on organisms of this type.

(iii) In addition, high counts foretell the likelihood of spoilage, because most foods contain 10^6 to 10^8 microorganisms per gram at the time when decomposition becomes evident (reviewed by Elliott and Michener, 1961).

With respect to (i) and (ii), the aerobic mesophilic bacteria generally (i.e., those that are recorded in plate counts made on bacterial culture media incubated at temperatures near 37° C) can be regarded as indicator organisms, though they are a much less precise and reliable measure of the hazard of food-poisoning than are the more closely related indicators which will be considered below.

The practice of using viable counts of aerobic instead of anaerobic bacteria probably developed because it was much easier to incubate under aerobic conditions, but recent developments, such as the use of deep agar in impermeable plastic bags (see, for example, Bladel and Greenberg, 1965), make anaerobic counting as easy and cheap as aerobic. However, there is no widespread experience to establish the significance of anaerobic counts. Anaerobic counts usually include many facultatively anaerobic organisms (such as coliforms, faecal streptococci, and staphylococci) unless special precautions are taken to exclude them, and more research is needed to establish the value of this kind of procedure. The anaerobic mesophilic count could serve as an indicator of conditions favourable to multiplication of anaerobic food-poisoning organisms such as *C. perfringens* and *C. botulinum*, and it is necessary when "screening" foods (especially those containing meat) where food-poisoning from *C. perfringens* is suspected.

In international trade, the importer has only the food itself for consideration, because he usually does not have extensive information on the conditions of sanitation or the time and temperature of its production and transport. Here an aerobic mesophilic count can be a helpful guide. If it is high, or if it varies widely among samples from different lots or within a lot, microbiological control in processing or transport was probably inadequate. The producer, on his part, can use such counts to test sanitation along his processing line. Samples are taken of the ingredients as they are added; and of the product, before and after those processing steps which may add or destroy microbes, and before, during, and after periods of delay which could permit growth to occur on or in the product.

There are important limitations to the value of any such count:

(i) In particular types of foods (e.g., some sausages, sauerkraut, cheese, and certain other dairy products) a large multiplication of bacteria with a concurrent "fermentation" or "ripening" of the food is desired. In these products a high count has practically no significance.

(ii) In heat-processed foods the viable count may be very low and actually reflects only the degree of heat treatment, but gross contamination of the raw material can sometimes be detected by a microscopic examination. (When pathogens are present, the absence of the normal spoilage flora may sometimes create a hazard. It has been shown that pathogens can often proliferate more easily when few spoilage organisms are present.)

(iii) Similarly, there is invariably a decrease in the number of viable bacteria when foods are dried or frozen. Thus, a plate count will not reflect the microbiological condition before processing, and direct microscopic examination is required to detect whether high counts occurred initially.

(iv) A mesophilic count is of little value in predicting the life of a food in chill storage, because many mesophilic microorganisms die at temperatures in the range 15° to 5° C or below (Michener and Elliott, 1964). For this purpose, a so-called psychrophilic viable count is preferred, usually at an incubation temperature in the range 0° to 5° C (Ingram, 1966).

Enteric indicator organisms

ESCHERICHIA COLI AND THE COLIFORMS (COLI-AEROGENES BACTERIA)

The native habitat for *Escherichia coli* is the enteric tract of man and other warm-blooded animals. Presence of this organism is generally considered to indicate pollution of direct or indirect faecal origin. Hence, *E. coli* is the classical indicator of the presence of enteric pathogens. Different pathogens may occur according to local disease rates in the production and processing areas: those of concern include *Salmonella typhi*, other salmonellae, shigellae, vibrios, entamoebae, various other zoonotic parasites, and enteric viruses. Many of these are difficult and costly to

determine, hence the common reliance upon the indicator species. However, it should be emphasized that the presence of *E. coli* has no direct
connotation of the presence of a pathogen; a risk only is implied.

Much has been published about the relative merits of *E. coli* and the
coliforms as faecal indicators (Buttiaux and Mossel, 1961; APHA, 1960,
1965, 1966; Hall, 1964; Lewis and Angelotti, 1964; Wilson and Miles,
1964; Geldreich, 1966). A common practice is to use tests for coliforms,
which include *E. coli*, for "screening" or preliminary tests. If there is
reason to determine the likelihood of faecal contamination, the coliforms
are subjected to further tests to establish whether any of them are *E. coli*.

The general term "coliforms" includes *E. coli* and species from other
genera of the Enterobacteriaceae; the taxonomic groups involved are
shown in Table 1. Coliforms other than *E. coli* persist in soil or on surfaces
(e.g., fruits, grain, field-crops) longer than will *E. coli*. Some species of
Erwinia, which closely resemble faecal coliforms, are plant pathogens
and do not indicate faecal contamination.

Thus, coliforms do not necessarily indicate contamination from a faecal

TABLE 1

Relation of the taxonomic groups of Enterobacteriaceae to their
occurrence and detection in foods

Group 1   Enteropathogenic types
          Salmonella
          Arizona
          Shigella
          Certain serotypes of *Escherichia coli*
          Certain serotypes of *Providencia*
Group 2   Possible indicators of faecal contamination
          Escherichia coli
          Proteus
          (To a lesser extent urease-positive biotypes of *Klebsiella* and
            Enterobacter aerogenes)
Group 3   Coli-aerogenes bacteria (fermenting lactose rapidly with
            formation of gas)
          Escherichia coli
          Citrobacter
          Enterobacter aerogenes
          Klebsiella
Group 4   Non-pathogenic bacteria generally not detected with coli-aerogenes tests
          Citrobacter
          Hafnia
          Certain biotypes of *Escherichia*
          Proteus
          Certain serotypes of *Providencia*
          Serratia
          Erwinia

NOTE *Enterobacter* is a newer name (Enterobacteriaceae Subcommittee, 1963) to
replace *Aerobacter*, which, however, is still widely used.

source, in the sense of implying immediate contact with faeces or a faecally contaminated surface. Further, the results of the quantitative determination of coliforms (including *E. coli*) in food may bear no relation to the extent of the original contamination where time-temperature abuses of the foods occur; this is in contrast to the quantitative results from clean water when numbers may be directly related to the degree of pollution. However, in a processed food, coliforms do indicate inadequate processing or post-processing contamination, most probably from workers, or from dirty instruments, machinery, or surfaces, or from the raw food before processing, which might derive their contamination from various sources (often human contact, polluted water, soil, or manure) depending on the food and the processing. The original contamination, whether via soil, water, or equipment, will frequently have been from faeces or sewage (*Erwinia* being excluded from this statement).

In foods, *E. coli* is the generally preferred indicator of contamination of relatively recent faecal origin, or from a substrate in which proliferation of *E. coli* (and associated pathogens, if present) has occurred. In appraising food safety, the presence of *E. coli* usually warrants a more positive assumption of hazard than the presence of other coliforms.

A newer term, "faecal coliforms," has arisen from attempts to find rapid dependable methods to establish the presence of *E. coli*, or closely related variants, without the need to purify the cultures or to apply the relatively costly IMViC tests. The faecal coliforms are a group of organisms selected by incubating inoculum derived from a coliform enrichment broth at temperatures higher than normal (44° to 45.5° C, depending on the method; see Part II, pp. 71, 77–79). Such preparations usually contain a high proportion of *E. coli* types I and II and are thus useful to indicate a probable faecal source (Fishbein, 1961; Fishbein and Surkiewicz, 1964; Geldreich, 1966; Guinée and Mossel, 1963).

As indicated above, coliforms not identified as *E. coli* may not indicate faecal contamination, and hence the potential presence of enteric pathogens. Also some of the more resistant coliforms, such as *Enterobacter*, may persist longer under unfavourable conditions than some enteric pathogens. However, the importance of the latter point remains problematic until adequate comparative data on longevity are available for the relevant enteric pathogens. Viable cells of *Salmonella typhi* can persist in cheese for months (Campbell and Gibbard, 1944), or in candy with a pH greater than 6 for a year or so (Bindseil, 1917; Kudicke, 1961), while *S. paratyphi* B has been recovered from frozen egg after several years of storage; other salmonellae may be found in dried foods from which large numbers of *E. coli* have been removed during processing.

The least debatable conclusions are these: (i) *E. coli* is a good indicator of faecal contamination, in most foods. (ii) Coliforms other than *E. coli* are a good indicator of unsatisfactory processing or sanitation; the presence of large numbers of coliforms in a processed food indicates that the opportunity for proliferation might have occurred, which could also have allowed multiplication of salmonellae, shigellae, staphylococci, or other organisms possibly introduced because of poor sanitation. (iii) The "faecal coliforms" have a higher probability of containing organisms of faecal origin and hence of indicating faecal contamination than do coliforms that have received no further differentiating tests.

TOTAL ENTEROBACTERIACEAE AS INDICATORS

A new development is the proposal (Mossel *et al.*, 1962, 1963) to use a medium which allows colony formation by all members of the Enterobacteriaceae, rather than to select as indicators only the lactose-positive members (e.g., *Escherichia* and *Enterobacter*). Various types of colony develop on the medium and, to determine the identity of a given colony and hence its probable origin, further identification procedures are required.

The advantages are the following: (i) It avoids misleading negative tests for coliforms in circumstances where pathogenic members of the family, such as *Salmonella*, may be present, as in foods containing processed egg. (ii) Several lactose-negative members of the family, such as *Salmonella* and *Shigella*, would not only indicate faecal contamination but would have more direct public health significance than the coliforms. (iii) In certain processes (such as radicidation[1] by ionizing radiation) some salmonellae may be more resistant than *Escherichia* or *Enterobacter* (*Aerobacter*); hence these common indicators would not reliably indicate the absence of *Salmonella* (Erdman *et al.*, 1961).

The method has been applied for various purposes such as testing the adequacy of sanitation of factory surfaces (Mossel *et al.*, 1966), and the estimation of the hygienic quality of components of animal feed (van Schothorst *et al.*, 1966) and of water used in food processing (von Buchli *et al.*, 1966). Laboratory collaboration between members of the Committee demonstrated that the medium for Enterobacteriaceae, when applied to specimens of meringue powder (essentially dried egg-white, sugar, and flavouring) with comparatively little contamination other than salmonellae, provided a "count" comparable to that provided by an effective

1/A treatment like pasteurization (Goresline *et al.*, 1964).

MPN (most probable number) procedure for salmonellae. This indicates that the test could be used to indicate possible enteric contamination in the absence of coliforms. Of course, additional tests are necessary to establish whether any of the developing colonies are *Salmonella*.

THE ENTEROCOCCI

Much has been written about the enterococci, and the Lancefield Group D streptococci generally, as faecal indicators. Group D includes, besides the enterococci, less resistant species such as *Strep. bovis* and *Strep. equinus*, the whole group being loosely called "faecal streptococci."

Although they are normally present in mammalian faeces, they have not been widely adopted as faecal indicators; because (e.g., in processed foods, water, and shellfish) there may be poor correlation between the presence of enterococci and of *E. coli* (APHA, 1958; Niven, 1963; Hartman *el al.*, 1965), although, on the other hand, in certain raw foods the correlation between the presence of enterococci and of coliforms is good (Simonsen, 1966). Buttiaux has drawn attention to the value, with raw foods, of the association between Group D streptococci and coli-aerogenes bacteria in giving indications of faecal contamination (Buttiaux, 1959; Buttiaux and Mossel, 1961).

Streptococcus faecium may predominate in faeces of several domestic animals while *Strep. faecalis* is normally important in the microflora of faeces of man. Hence, *Strep. faecium* was suggested as indicating faecal contamination originating in domestic animals rather than man (Barnes and Ingram, 1956). However, additional study has established the uncertainty of this specificity and of the relative proportion between the two in a given host (Barnes, 1964). Organisms classified as *Strep. faecalis*, *Strep. faecium*, and *Strep. bovis* have all been found in faeces of man and of cattle and pigs (Raibaud *et al.*, 1961; Tilton and Litsky, 1967), and as *Strep. faecalis* and *Strep. bovis* in sheep. All have been found in frozen pot pies (Tilton and Litsky, 1967). Further, the reliability of simple determinative methods is open to question (Deibel, 1964; Tilton and Litsky, 1967), though the method of Barnes (1956) has been widely used in Europe instead of the procedures based on azide (cf. Part II, pp. 86–87).

Nevertheless, large numbers of enterococci in a food do indicate doubtful microbiological quality (except where specific strains have been used to ferment the food), because their presence implies either exposure to conditions which might permit extensive multiplication of many undesirable species or inadequate sanitary practice.

The enterococci may have a distinctive role as indicators of poor factory sanitation, owing to their relatively high resistance to drying, high temperature, detergents, or disinfectants. Because of their greater resistance to freezing, the enterococci are sometimes the preferred indicator of poor plant sanitation with frozen foods (Niven, 1963; Simonsen, 1966), and, similarly, they may be informative with foods which have been heated or dehydrated with consequent destruction of more labile indicators such as the Enterobacteriaceae.

This high resistance is, at the same time, the reason for the unreliability of these cocci as general indicators of faecal contamination. They may survive adverse conditions, so that they have little relation to the hazard from the less durable pathogens such as *Salmonella* or *Shigella* which, even if deposited simultaneously with the cocci, would probably not have survived.

The Committee offers no suggested limits of tolerance for the enterococci. Selective judgment based on experience is essential to interpret the significance of specific numbers in a particular food. The faecal streptococci are discussed in this context by Niven (1963), Lewis and Angelotti (1964), and Hartman *et al.* (1965).

Other indicator organisms

Other organisms occasionally cited as indicators include *Streptococcus salivarius*, for contamination from oral sources, and *Staphylococcus aureus*, for contamination from the mouth, nose, or skin. These particular organisms are not widely used as indicators, probably because each has certain disadvantages.

STREPTOCOCCUS SALIVARIUS

Streptococcus salivarius has been used as an indicator of oral contamination in food factories (Thatcher, 1955). Because this organism is almost always present in the human mouth, it becomes widely distributed on surfaces near workers who cough or spit. The organism dies rapidly on dry surfaces. After drying on several types of surface, and using Mitis-Salivarius agar (Chapman, 1944), heavy inocula were not recovered after a 24-hour exposure at room temperature (Thatcher, unpublished data). The demonstration of *Strep. salivarius* on factory surfaces thus indicates a lapse in worker hygiene less than 24 hours before making the test. Mitis-

Salivarius agar permits the separation from enterococci on the basis of colony characteristics and is useful for this purpose even in the presence of gross contamination. Confirmatory tests, if considered necessary, are available from Williams *et al.* (1956) and Cowan and Steel (1965).

STAPHYLOCOCCI

The presence of *Staph. aureus* in a food is usually taken to indicate contamination from the skin, mouth, or nose of workers handling the food, but inadequately cleaned equipment may also be a source of contamination. The presence of large numbers of staphylococci is, in general, a good indication that sanitation and temperature control have somewhere been inadequate.

Ingram (1960) attached significance to the presence of high numbers of any kind of *Staphylococcus* (mesophilic facultatively anaerobic catalase-positive micrococci), for example, in cured meats, because, whenever counts of this whole group are high, there are circumstances in which pathogenic strains of *Staph. aureus* might apparently occur in dangerous numbers. However, the development of specific methods for detecting coagulase-positive *Staph. aureus* has made the use of indicators less necessary.

CLOSTRIDIA

In canned non-acid foods, the presence of mesophilic clostridia probably indicates that heating was insufficient to destroy spores of *C. botulinum* which might have been present.

In chilled or dried foods, the presence of clostridia in unusually high numbers, or as an unusually large proportion of the total bacterial population, could indicate a hazard from *C. perfringens* or *C. botulinum* either in the food as prepared or in its future use.

THE IMPORTANCE OF DETECTION

Contamination in factory-processed foods

Many factors influence the kind and number of microorganisms present in a processed food. These include: (a) the initial contamination of the raw material upon arrival at the food-processing factory; (b) the additional contamination gained within the plant from several potential sources, such as manual contact, human fomites of all types, contact with surfaces (including machinery), air-borne and water-borne contamination, and the use of contaminated accessory ingredients or food additives; (c) the temperature of the food during storage and pre-processing; (d) the duration of exposure at temperatures that allow proliferation; (e) the severity of the process in terms of lethality to organisms of varying resistance; (f) post-processing contamination, which can include all the sources listed in (b) above, together with additional risks from sources involved in accessory procedures such as bottling, packaging, wrapping, and also cooling, whether by use of water, as in canning, or by air as in blast freezing or other cooling systems; (g) the selective effect of the process on the ecology of a surviving population.

The overall effects of contaminating factors will determine the quality of the food, its probable shelf-life, and the potential public health risks. Through microbiological examination several deductions can be made bearing on health hazards, particularly in foods which have been subjected to a process that will have destroyed specific types of microorganisms. For example, normal cooking, pasteurization, spray-drying, and similar applications of heat will have destroyed, in the treated foods, most members of the Enterobacteriaceae, Pseudomonadaceae, diphtheroids (Corynebacteria), lactobacilli, staphylococci, and non-thermoduric micrococci and streptococci. Again, properly conducted thermal canning will have destroyed spores of *Clostridium botulinum*, and of all but the most heat-resistant of other spore-bearing species, except where the heat process is partially dependent upon inhibitory "curing" agents, such as sodium nitrite and sodium chloride, in which case viable spores may be found, even though the outgrowth of spores is restricted by the presence of sufficient curing salts. Thus, if microbiological examination reveals the presence in a food of organisms of a type that supposedly should have been destroyed, the indication is that usually either the processing was inadequate or there has been post-processing contamination. It is generally recognized that the latter type of contamination can be potentially the more dangerous because, compared with raw foods, many processed foods are more likely to provide a suitable medium for proliferation of pathogens and toxinogenic bacteria. Also, since contamination can be held to a minimum under conditions of effective sanitation and worker hygiene, unnecessary contamination denotes failure in these respects, with a consequent increase in the risk of hazardous contamination from nasal, suppurative, expectorative, and faecal sources, and by cross-contamination from raw to processed products. The risk is increased still more because unsanitary factories and machinery encourage the proliferation of hazardous bacteria, including pathogens from human or animal sources and spore-formers, including *C. botulinum*, from environmental sources.

Exposure of non-sterile foods to temperatures which allow the proliferation of contaminants is of course a critical factor in the shelf-life and safety of such foods. When it is possible to enumerate in foods certain species, or groups of species, which grow under particular environmental conditions, much may be deduced about the probable source of contamination of the food, and sometimes the probable disease hazard entailed in consumption of these foods can be assessed. In attempting such assessments, enquiry is generally directed to the presence of known infective pathogens, the food-poisoning organisms, the so-called indicator organisms, or the total number of organisms that will grow on media of defined composition when incubated under defined environmental conditions.

Interpretation of the relative importance of pathogens in foods

The presence of pathogens in any foodstuff is to be avoided by all possible means. The Committee has already recorded its opinion that principles of public health require that no tolerance should be established for significant infectious pathogens. The Committee also recognizes that this policy may sometimes be difficult to implement owing to problems in technical feasibility, to sampling methods, and to economic and political pressures (Thatcher, 1963). What constitutes a significant pathogen is not always obvious, and judgment based on experience must often be exercised. The reader is referred to the discussion on *Staphylococcus* (pp. 17–18), where a distinction is made between the roles of an organism which, in different circumstances, may be an important pathogen, a source of food-poisoning toxins, an indicator of unsanitary practices, or a harmless contaminant of no concern.

Even though no formal tolerance for pathogens should be recognized in food-control legislation, the mere finding of particular toxinogenic organisms in foods (as distinct from infective pathogens) should not necessarily warrant removal of such foods from the market. Judgment based on experience must again be invoked. For example, milk may become contaminated with soil or manure, and hence with spores of *Clostridium botulinum*. The processing of such milk for marketing either as powdered milk or as cheese would probably not destroy such spores, but in the powdered milk the spores would not germinate and hence could not form toxin, while in "natural" cheese, the presence of many other competitive organisms, particularly the acid-forming lactic streptococci used as "starter cultures," either prevents the botulinum organisms from proliferating to the extent necessary to produce injurious amounts of toxin, or else the toxin, if formed, will be destroyed by other organisms before it becomes detectable (Thatcher, 1955; Wagenaar and Dack, 1955, 1958).

On the other hand, if such foods were to be added as ingredients of non-acid canned foods, it would be essential that the complete food mixture, unless protected by curing salts, receive a heat process sufficient to destroy all spores of *C. botulinum*. Otherwise, germination of spores and toxin formation might occur and give rise to botulism.

Thus, the importance of toxinogenic organisms in a food is influenced by the nature and formulation of the food, as this may influence proliferation, and by the anticipated treatment to be given the food. This has been discussed in greater detail elsewhere by the Committee (Thatcher, 1963).

Infective enteric pathogens have no place in foods. Infection may result from direct consumption of the food or indirectly through secondary contamination of other foods, which can occur by many devious methods. This applies particularly to *Salmonella typhi*, to other salmonellae, to shigellae, to enteric viruses including the virus of infectious hepatitis, and to many zoonotic parasites.

Research needs

At the time of writing, the Committee recognizes the need for further research in many subjects before reliable methods and criteria can be selected for the evaluation of all significant microorganisms or toxins in foods in international trade. Examples are given below, without any statement as to their relative importance, which varies with time and country.

Adequate techniques are not available for the isolation of viruses from foods. The available techniques are insensitive, require special facilities, and may involve prolonged observation of tissue cultures or test animals. For these reasons, there is little information on the range of viruses that may be transmitted through foods, or on the role of particular foods as vehicles of infection. The survival of viruses in foods is largely unknown, as well as their destruction during processes such as heating, freezing, desiccation, and irradiation.

Certain fungal toxins, notably the aflatoxins, have been shown to cause cancer in different species of animals, besides being implicated in other serious diseases. More direct evidence is needed to decide whether mycotoxins may cause similar effects in humans, and to extend knowledge about the kinds of fungi and environmental factors favouring their elaboration.

In some regions of the world, parasitic diseases such as amoebiasis, anisakiasis, hydatidosis, sarcosporidiosis, and toxoplasmosis occur, but little attention is usually given to their detection in the food laboratory.

Even among the well-known enteric bacteria, methods of detection in foods are poorly developed for some genera, especially *Shigella*. Although serological techniques have been useful for identifying enteropathogenic strains of *Escherichia coli*, these procedures have not been widely used by food microbiologists.

Vibrio parahaemolyticus, as has been stated, is an important food-borne pathogen associated with raw sea-foods in Japan. There is no apparent reason why its occurrence should be limited to Asiatic Pacific areas, and studies to establish its true distribution should be undertaken.

Large numbers of bacteria, of various species not usually regarded as pathogenic, are suspected of causing outbreaks of undetermined aetiology. Certain strains of aerobic bacilli, faecal streptococci, *Proteus* sp., and pseudomonads have been implicated on occasion. In several countries such as the United Kingdom, the United States, and Japan, nearly half the reported outbreaks of food-borne disease are of unknown aetiology and may well be due to some of the agents just mentioned.

Clostridium botulinum has been known for many years, but research has been stimulated recently by outbreaks of type E *C. botulinum*. The fact that type E produces thermolabile spores focuses attention on the weaknesses of established methods of isolating *C. botulinum*, which usually involve heating the samples. More rapid and reliable methods are desirable for isolating this organism and for detecting specific types of its toxin. Extensive studies are in progress along these lines.

It is not yet known whether those strains of *Clostridium perfringens* implicated in food-poisoning belong to an identifiable group, and the mechanism by which *C. perfringens* causes diarrhoea is not yet clear.

Although a number of useful procedures have been devised for enumerating staphylococci in foods, more information is needed regarding the influence of the concomitant flora and of various food constituents on the growth and colonial appearance of toxinogenic strains. Conditions affecting the production and activation or inactivation of enterotoxins are incompletely understood. The most critical need is to establish how many types of enterotoxin may be elaborated in foods and to provide specific antisera for their detection in food extracts.

The development of a reliable and rapid procedure for the detection of salmonellae in foods is urgently needed to assist quality control laboratories and regulatory agencies in examining the large number of samples associated with commercial food-processing operations. Methods for the isolation of salmonellae are still diffuse, and clarification is needed on several points, such as the relative application of non-selective and selective enrichment, and temperature and time of incubation. The influence of ill-defined variations in media on their effectiveness for the isolation of salmonellae needs intensive study. More extensive application of typing phages is needed for identification of commonly occurring salmonellae. Further testing of a recently described phage, which presumably affects all salmonellae, is much needed.

Also, in general, the effect of the food substance on the selectivity and nutritional quality of culture media should be examined further. Differences among media and media ingredients, as well as variations among commercial lots of dehydrated products, are major barriers to the standardization of microbiological methods on an international basis.

REFERENCES

AISO, K. and MATSUNO, M. 1961. The outbreaks of enteritis-type food poisoning due to fish in Japan and its causative bacteria. Japan J. Microbiol. 5, 337.

AMERICAN PUBLIC HEALTH ASSOCIATION. 1958. Recommended methods for the microbiological examination of foods (1st ed.; New York: American Public Health Association, Inc.).

——— 1960. Standard methods for the examination of dairy products (11th ed.; New York: American Public Health Association, Inc.).

AMERICAN PUBLIC HEALTH ASSOCIATION, AMERICAN WATER WORKS ASSOCIATION, and WATER POLLUTION CONTROL FEDERATION. 1965. Standard methods for the examination of water and wastewater (12th ed.; New York: American Public Health Association, Inc.).

AMERICAN PUBLIC HEALTH ASSOCIATION, SUBCOMMITTEE ON METHODS FOR THE MICROBIOLOGICAL EXAMINATION OF FOODS. 1966. Recommended methods for the microbiological examination of foods (2nd ed.; New York: American Public Health Association, Inc.).

ANDERSON, P. H. R. and STONE, D. M. 1955. Staphylococcal food poisoning associated with spray-dried milk. J. Hyg., Camb. 53, 387.

ANGELOTTI, R., HALL, H. E., FOTER, M. J., and LEWIS, K. H. 1962. Quantitation of Clostridium perfringens in foods. Appl. Microbiol. 10, 193.

ARMIJO, R., HENDERSON, D. A., TIMOTHEE, R., and ROBINSON, H. B. 1957. Food poisoning outbreaks associated with spray-dried milk—An epidemiologic study. Am. J. Public Health 47, 1093.

BARNES, E. M. 1956. Tetrazolium reduction as a means of differentiating Streptococcus faecalis from Streptococcus faecium. J. Gen. Microbiol. 14, 57.

——— 1964. Distribution and properties of serological types of Streptococcus durans and related strains. J. Appl. Bacteriol. 27, 461.

BARNES, E. M. and INGRAM, M. 1956. The distribution and significance of different species of faecal streptococci in bacon factories. J. Appl. Bacteriol. 19, 204.

BEERENS, H. and TAHON-CASTEL, M. 1965. Infections humaines à bactéries, anaérobies non toxigènes, Vol. 1 (Brussels: Presses Académiques Européennes).

BERGDOLL, M. S. 1963. The nature and detection of staphylococcal enterotoxin. In Microbiological quality of foods, Proc. of Conf., Franconia, N. H., 27–29 Aug. 1962 (New York: Academic Press), Discussion, p. 54.

BERGDOLL, M. S., SUGIYAMA, H., and DACK, G. M. 1959. Staphylococcal enterotoxin. I. Purification. Arch. Biochem. Biophys. 85, 62.

BERGDOLL, M. S., WEISS, K. F., and MUSTER, M. J. 1967. The production of staphylococcal enterotoxin by a coagulase-negative microorganism. Bacteriol. Proc. *67*, 12.

BINDSEIL, F. 1917. Über die Haltbarkeit der Typhusbazillen an Nahrungs- und Genussmitteln. Z. Hyg. Infektionskr. *84*, 181.

BLADEL, B. O. and GREENBERG, R. A. 1965. Pouch method for the isolation and enumeration of clostridia. Appl. Microbiol. *13*, 281.

BODNAR, S. 1962. Über durch *Bac. cereus* verursachte alimentäre atypisch verlaufende Lebensmittelvergiftungen. Z. ges. Hyg. Grenzgeb. *8*, 388.

BORJA, C. R. and BERGDOLL, M. S. 1967. Purification and partial characterization of enterotoxin C produced by *Staphylococcus aureus* strain 137. Biochem. J. *6*, 1467.

BOWMER, E. J. 1964. The challenge of salmonellosis: Major public health problem. Am. J. Med. Sci. *274*, 467.

BUCHBINDER, L., OSLER, A. G., and STEFFEN, G. I. 1948. Studies in enterococcal food poisoning. I. The isolation of enterococci from foods implicated in several outbreaks of food poisoning. Public Health Rept. (U.S.) *63*, 109.

BUTTIAUX, R. 1959. The value of the association Escherichieae–Group D streptococci in the diagnosis of contamination in foods. J. Appl. Bacteriol. *22*, 153.

BUTTIAUX, R. and MOSSEL, D. A. A. 1961. The significance of various organisms of faecal origin in foods and drinking water. J. Appl. Bacteriol. *24*, 353.

CAMPBELL, A. G. and GIBBARD, J. 1944. The survival of *E. typhosa* in cheddar cheese manufactured from infected raw milk. Can. J. Public Health *35*, 158.

CASMAN, E. P. 1966. Staphylococcal food poisoning. Presented at the 94th Annual Meeting, Amer. Public Health Assoc., San Francisco, Calif., Oct. 31 to Nov. 4.

CASMAN, E. P. and BENNETT, R. W. 1965. Detection of staphylococcal enterotoxin in food. Appl. Microbiol. *13*, 181.

CASMAN, E. P., BENNETT, R. W., DORSEY, A. E., and ISSA, J. A. 1967. Identification of a fourth staphylococcal enterotoxin, Enterotoxin D. J. Bacteriol. *94*, 1875.

CASMAN, E. P., BERGDOLL, M. S., and ROBINSON, J. 1963. Designation of staphylococcal enterotoxins. J. Bacteriol. *85*, 715.

CHAPMAN, G. H. 1944. The isolation of streptococci from mixed cultures. J. Bacteriol. *48*, 113.

CHRISTIANSEN, O., KOCH, S. O., and MADELUNG, P. 1951. Et utbrud af levnedsmiddelforgiftning forårsaget af *Bacillus cereus*. Nord. Veterinarmed. *3*, 194.

CLARENBURG, A. and KAMPELMACHER, E. H. 1957. *Bacillus cereus* als oorzaak van voedselvergiftiging. Voeding *18*, 384.

CLEERE, R. L., MOLLOHAN, C. S., and REID, G. 1967. Shigellosis in Denver, Colorado. An investigation of a possible relationship between eggs and shigellosis. Shigella Surveillance Report No. 14, 3.

COSTIN, I., DAVID, P., DINCULESCU, M., WEISZBERGER, A., OLARIU, GH., SCHIFTEN, N., and MARTON, T. 1960. Toxiinfectie alimentara cu germeni din genul *Escherichia*. Microbiol. Parazitol. Epidemiol. (Bucharest) *6*, 531.

COSTIN, I. D., VOICULESCU, D., and GORCEA, V. 1964. An outbreak of food poisoning in adults associated with *Escherichia coli* serotype 86: B_7H_{34}. Pathol. Microbiol. 27.

COWAN, S. T. and STEEL, K. J. 1965. Manual for the identification of medical bacteria (Cambridge: Cambridge University Press), p. 56.

CRUICKSHANK, R. 1965. Medical microbiology (11th ed.; Edinburgh: E & S Livingstone Ltd.).

DACK, G. M. 1956. Food poisoning (3rd ed.; Chicago: University of Chicago Press).

DEIBEL, R. H. 1964. The Group D streptococci. Bacteriol. Rev. 28, 330.

DICKIE, N. and THATCHER, F. S. 1967. Severe food-poisoning from malted milk and the isolation of a hitherto unknown toxin (Abstr.) Can. J. Public Health 58, 25.

DISCHE, F. E. and ELEK, S. D. 1957. Experimental food-poisoning by *Clostridium welchii*. Lancet 2, 71.

DOLMAN, C. E. 1942. Staphylococcus enterotoxin. Proc. 6th Pacific Science Congress, 1939, 5, 363.

―――― 1943. Bacterial food poisoning. Can. J. Public Health 34, 205.

―――― 1957. The epidemiology of meat-borne diseases. *In* Meat hygiene. (World Health Organization, Monograph Series, No. 33).

DRACHMAN, R. H., PAYNE, F. J., JENKINS, A. A., MACKEL, D. C., PETERSEN, N. J., BORING, J. R., GAREAU, F. E., FRASER, R. S., and MYERS, G. G. 1960. An outbreak of water-borne *Shigella* gastroenteritis. Am. J. Hyg. 72, 321.

EDWARDS, P. R. and EWING, W. H. 1962. Identification of *Enterobacteriaceae* (2nd ed.; Minneapolis: Burgess Pub. Co.).

EDWARDS, P. R. and GALTON, M. M. 1967. *Salmonellosis*. Advan. Vet. Sci. 11, 1.

EDWARDS, P. R., GALTON, M. M., BRACHMAN, P. S., and MCCALL, C. E. 1964. A perspective of salmonellosis (Atlanta, Ga.: Communicable Disease Center, U.S. Department of Health, Education and Welfare, Public Health Service).

EDWARDS, P. R., MCWHORTER, A. C., and FIFE, M. A. 1956. The Arizona group of *Enterobacteriaceae* in animals and man. Occurrence and distribution. Bull. World Health Org. 14, 511.

ELLIOTT, R. P. and MICHENER, H. D. 1961. Microbiological standards and handling codes for chilled and frozen foods. Appl. Microbiol. 9, 452.

ENTEROBACTERIACEAE SUBCOMMITTEE OF THE NOMENCLATURE COMMITTEE OF THE INTERNATIONAL ASSOCIATION OF MICROBIOLOGICAL SOCIETIES. 1963. Report of 1962 Meeting, Montreal. Int. Bull. Bacteriol. Nomencl. Taxon. 13, 141.

ERDMAN, I. E., THATCHER, F. S., and MacQUEEN, K. F. 1961. Studies on the irradiation of microorganisms in relation to food preservation. I. The comparative sensitivities of specific bacteria of public health significance. Can. J. Microbiol. 7, 199.

EWING, W. H., DAVIS, B. R., and MONTAGUE, T. S. 1963. Studies on the occurrence of *Escherichia coli* serotypes associated with diarrhoeal diseases (Atlanta, Ga.: Communicable Disease Center, Laboratory Branch, U.S. Department of Health, Education and Welfare).

EWING, W. H., TATUM, N. W., and DAVIS, B. R. 1957. The occurrence of *Escherichia coli* serotypes associated with diarrhoeal disease in the United States. Public Health Lab. 15, 118.

FILIPPONE, M. V., MITCHELL, I. A., BRAYTON, J. B., NEWELL, K. W., and SMITH, M. H. D. 1967. Comparison of fluorescent antibody with cultural technique for isolation of enteropathogenic *Escherichia coli* from swine. Appl. Microbiol. *15*, 1437.

FISHBEIN, M. 1961. The aerogenic response of *Escherichia coli* and strains of *Aerobacter* in EC broth and selected sugar broths at elevated temperatures. Appl. Microbiol. *10*, 79.

FISHBEIN, M. and SURKIEWICZ, B. F. 1964. Comparison of the recovery of *Escherichia coli* from frozen foods and nutmeats by confirmatory incubation in EC medium at 44.5 and 45.5° C. Appl. Microbiol. *12*, 127.

FOOD PROTECTION COMMITTEE, U.S. NATL. ACAD. SCI. – NATL. RES. COUNCIL. 1964. An evaluation of public health hazards from microbiological contamination of foods. A Report: Publication 1195 (Washington, D.C.: N.A.S.–N.R.C.).

FUJINO, T., OKUNO, Y., NAKADA, D., AOYAMA, A., FUKAI, K., MUKAI, T., and UEHO, T. 1953. On the bacteriological examination of shirasu food poisoning. Med. J. Osaka Univ. *4*, 299.

GELDREICH, E. E. 1966. Sanitary significance of fecal coliforms in the environment (U.S. Department of the Interior, Federal Water Pollution Control Administration, Washington, D.C., Publ. WP-20-3).

GEORGALA, D. L. and BOOTHROYD, M. 1964. A rapid immunofluorescence technique for detecting salmonellae in raw meat. J. Hyg., Camb. *62*, 319.

GORESLINE, H. E., INGRAM, M., MACÚCH, P., MOCQUOT, G., MOSSEL, D. A. A., NIVEN, C. F., JR., and THATCHER, F. S. 1964. Tentative classification of food irradiation processes with microbiological objectives. Nature *204*, 237.

GUINÉE, P. A. M. and MOSSEL, D. A. A. 1963. The reliability of the test of McKenzie, Taylor and Gilbert for detection of faecal *Escherichia coli* strains of animal origin in foods. Antonie van Leeuwenhoek, J. Microbiol. Serol. *29*, 163.

HAGLUND, J. R., AYRES, J. C., PATON, A. M., KRAFT, A. A., and QUINN, L. Y. 1964. Detection of Salmonella in eggs and egg products with fluorescent antibody. Appl. Microbiol. *12*, 447.

HALL, H. E. 1964. Methods of isolation and enumeration of coliform organisms. *In* Lewis and Angelotti (1964).

HALL, H. E., ANGELOTTI, R., LEWIS, K. H., and FOTER, M. J. 1963. Characteristics of *Clostridium perfringens* strains associated with food and food-borne disease. J. Bacteriol. *85*, 1094.

HALL, H. E., BROWN, D. F., and LEWIS, K. H. 1967. Examination of market foods for coliform organisms. Appl. Microbiol. *15*, 1062.

HALL, H. E. and HAUSER, G. H. 1966. Examination of feces from food handlers for salmonellae, shigellae; enteropathogenic *Escherichia coli*, and *Clostridium perfringens*. Appl. Microbiol. *14*, 928.

HARTMAN, P. A., REINBOLD, G. W., and SARASWAT, D. C. 1965. Indicator organisms, a review: Role of enterococci in food-poisoning. J. Milk Food Technol. *28*, 344.

HARTSELL, S. F. 1951. The longevity and behaviour of pathogenic bacteria in frozen foods: The influence of plating media. Am. J. Public Health *41*, 1072.

HAUGE, S. 1955. Food poisoning caused by aerobic spore-forming bacilli. J. Appl. Bacteriol. *18*, 591.

HAUSCHILD, A. H. W., ERDMAN, I. E., HILSHEIMER, R., and THATCHER, F. S. 1967a. Variations in recovery of *Clostridium perfringens* on commercial sulfite-polymyxin-sulfadiazine (SPS) agar. J. Food Sci. *32*, 469.

HAUSCHILD, A. H. W., HILSHEIMER, R., and THATCHER, F. S. 1967b. Acid resistance and infectivity of food-poisoning *Clostridium perfringens*. Can. J. Microbiol. *13*, 1041.

HAUSCHILD, A. H. W., NIILO, L., and DORWARD, W. J. 1967c. Experimental enteritis with food-poisoning and "classical" strains of *Clostridium perfringens* type A in lambs. J. Infect. Diseases *117*, 379.

HAUSCHILD, A. H. W. and THATCHER, F. S. 1967. Experimental food-poisoning with heat-susceptible *Clostridium perfringens* type A. J. Food Sci. *32*, 467.

—— 1968. Experimental gas gangrene with food-poisoning *Clostridium perfringens* type A. Can. J. Microbiol. *14*, 705.

HOBBS, B. C. 1965a. Contamination of meat supplies. Mon. Bull. Minist. Health Lab. Serv. *24*, 123 and 145.

—— 1965b. *Clostridium welchii* as a food poisoning organism. J. Appl. Bacteriol. *28*, 74.

HOBBS, B. C., SMITH, M. E., OAKLEY, C. L., WARRACK, G. H., and CRUICKSHANK, J. C. 1953. *Clostridium welchii* food poisoning. J. Hyg., Camb. *51*, 75.

INGRAM, M. 1960. Bacterial multiplication in packed Wiltshire bacon. J. Appl. Bacteriol. *23*, 206.

—— 1966. Psychrophilic and psychrotrophic microorganisms. Ann. Inst. Pasteur, Lille *15*, 111.

INGRAM, M. and ROBERTS, T. A. (eds.). 1966. Botulism 1966: Proc. 5th Intern. Symp. on Food Microbiology, Moscow (London: Chapman and Hall, 1967).

INSALATA, N. F., SCHULTE, S. J., and BERMAN, J. H. 1967. Immunofluorescence technique for detection of salmonellae in various foods. Appl. Microbiol. *15*, 1145.

JOHNSON, H. M., HALL, H. E., and SIMON, M. 1967. Enterotoxin B: Serological assay in cultures by passive haemagglutination. Appl. Microbiol. *15*, 815.

JOINT WHO/FAO EXPERT COMMITTEE ON ZOONOSES. 1959. Second Report (World Health Organization, Technical Report Series, No. 169).

JOSEPH, R. L. and BAIRD-PARKER, A. C. 1965. Fractions of staphylococcal enterotoxin B. Nature *207*, 663.

KAISER, R. L. and WILLIAMS, L. D. 1962. Trace two bacillary dysentery outbreaks to single food source. Penn. Med. J. *65*, 351.

KAWABATA, T. and SAKAGUCHI, G. 1963. Halophilic bacteria as a cause of food poisoning. *In* Microbiological quality of foods, Proc. of Conf., Franconia, N.H., 27–29 Aug. 1962 (New York: Academic Press).

KELLER, M. D. and ROBBINS, M. L. 1956. An outbreak of *Shigella* gastroenteritis. Public Health Rept. *71*, 856.

KORETSKAIA, L. S. and KOVALEVSKAIA, A. N. 1958. Food poisoning produced by *B. coli* serotype $O_{26}B_6$. Zh. Mikrobiol., Epidemiol. i Immunobiol. *4*, 58 (Eng. transl. in J. Microbiol., Epidemiol., Immunobiol. (USSR) *29*, 553).

KUDICKE, H. 1961. Ueber die Lebensdauer von *Salmonella*-Bakterien auf Süsswaren. Oeffentl. Gesundheitsdienst *23*, 109.

LEVY, E., RIPPON, J. E., and WILLIAMS, R. E. O. 1953. Relation of bacteriophage pattern to some biological properties of staphylococci. J. Gen. Microbiol. 9, 97.

LEWIS, K. H. and ANGELOTTI, R. (eds.). 1964. Examination of foods for entero-pathogenic and indicator bacteria. Review of methodology and manual of selected procedures (Division of Environmental Engineering and Food Protection, U.S. Public Health Service, Publ. No. 1142).

LEWIS, K. H. and CASSEL, K., JR. (eds.). 1964. Botulism. Proc. of Symp., Cincinnati, Ohio, Jan. 13–15, 1964 (U.S. Public Health Service Publ. No. 999-FP-1 [Cincinnati, Ohio: Public Health Service]).

MCCLUNG, L. S. 1945. Human food-poisoning due to growth of *Clostridium perfringens* (*C. welchii*) in freshly cooked chicken: Preliminary note. J. Bacteriol. 50, 229.

MEYER, K. F. 1953. Food poisoning. New Engl. J. Med. 249, 765.

MICHENER, H. D. and ELLIOTT, R. P. 1964. Minimum growth temperatures for food-poisoning, fecal-indicator, and psychrophilic microorganisms. Advan. Food Res. 13, 349.

MINISTRY OF AGRICULTURE, FISHERIES AND FOOD, THE VETERINARY LABORATORY SERVICE, and THE PUBLIC HEALTH LABORATORY SERVICE. 1965. Salmonellae in cattle and their feeding stuffs, and the relation to human infection. J. Hyg., Camb. 63, 223.

MIYAMOTO, Y., NAKAMUMA, K., and TAKIZAWA, K. 1961. Pathogenic halophiles. Proposal of a new Genus "Oceanomonas" and of the amended species names. Japan J. Microbiol. 5, 477.

MORSE, S. A. and MAH, R. A. 1967. Microtiter Haemagglutination-Inhibition Assay for staphylococcal enterotoxin B. Appl. Microbiol. 15, 58.

MOSSEL, D. A. A. 1963. La survie des salmonellae dans les différents produits alimentaires. Ann. Inst. Pasteur 104, 551.

MOSSEL, D. A. A., BEERENS, H., TAHON-CASTEL, M., and POLSPOEL, B. 1965. Etudes des milieux utiles pour le dénombrement des spores des bactéries anaérobies en microbiologie alimentaire. Ann. Inst. Pasteur, Lille 15, 147.

MOSSEL, D. A. A., KAMPELMACHER, E. H., and VAN NOORLE-JANSEN, L. M. 1966. Verification of adequate sanitation of wooden surfaces used in meat and poultry processing. Zentr. Bakteriol., Parasitol., Infekt. Hyg. 201, 91.

MOSSEL, D. A. A., MENGERINK, W. H. J., and SCHOLTS, H. H. 1962. Use of a modified MacConkey agar medium for the selective growth and enumeration of *Enterobacteriaceae*. J. Bacteriol. 84, 381.

MOSSEL, D. A. A., VISSER, M., and CORNELISSEN, A. M. R. 1963. The examination of foods for *Enterobacteriaceae* using a test of the type generally adopted for the detection of salmonellae. J. Appl. Bacteriol. 26, 444.

MUNCH-PETERSEN, E. 1960. Food-borne epidemics due to staphylococci. Australian J. Dairy Technol. 15, 25.

MUNCH-PETERSEN, E. and BOUNDY, C. 1962. Yearly incidence of penicillin-resistant staphylococci in man since 1942. Bull. World Health Org. 26, 241.

NAKAMURA, M. and DAWSON, D. A. 1962. Role of suspending and recovery media in the survival of frozen *Shigella sonnei*. Appl. Microbiol. 10, 40.

NETER, E., WEBB, C. R., SHUMWAY, C. N., and MURDOCK, M. R. 1951. Study on etiology, epidemiology and antibiotic therapy of infantile diarrhea, with a particular reference to certain serotypes of *Escherichia coli*. Am. J. Public Health *41*, 1490.

NEWELL, K. W. 1959. The investigation and control of salmonellosis. Bull. World Health Org. *21*, 279.

NEWELL, K. W., SMITH, M. H. D., BRAYTON, J. B., MITCHELL, I., and RODER, R. 1966. The zoonotic hypothesis for human enteropathogenic *Escherichia coli* infection. Presented to the Epidemiology Section, Amer. Public Health Assoc., Francisco, Calif., Nov.

NIKODEMUSZ, I. 1958. *Bacillus cereus* als Ursache von Lebensmittelvergiftungen. Z. Hyg. Infektionskr. *145*, 335.

NIVEN, C. F., JR. 1963. Microbial indexes of food quality: Fecal streptococci. p. 119. *In* Microbial quality of foods, Proc. Conf., Franconia, N.H., 27–29 Aug. 1962 (New York: Academic Press).

NYGREN, B. 1962. Phospholipase C-producing bacteria and food poisoning. An experimental study on *Clostridium perfringens* and *Bacillus cereus*. Acta Pathol. Microbiol. Scand. Suppl. 160.

PARKER, M. T. and LAPAGE, S. P. 1957. Penicillinase production by *Staphylococcus aureus* strains from outbreaks of food poisoning. J. Clin. Pathol. *10*, 313.

RAIBAUD, P., CAULET, M., GALPIN, J. V., and MOCQUOT, G. 1961. Studies of the bacterial flora of the alimentary tract of pigs. II. Streptococci: Selective enumeration and differentiation of the dominant group. J. Appl. Bacteriol. *24*, 285.

SAKAZAKI, R., IWANAMI, S., and FUKUMI, H. 1963. Studies on the enteropathogenic, facultatively halophilic bacteria, *Vibrio parahaemolyticus*. I. Japan J. Med. Sci. Biol. *16*, 161.

SALMONELLA SURVEILLANCE. 1964 *et seq.* Reports issued monthly by the National Communicable Disease Center, U.S. Public Health Service. (Salmonella surveillance; annual summary, 1963—available from the Center, Atlanta, Ga.)

SHELTON, L. R., LEININGER, H. V., SURKIEWICZ, B. F., BAER, E. F., ELLIOTT, R. P., HYNDMAN, J. B., and KRAMER, N. 1962. A bacteriological survey of the frozen precooked food industry (Washington, D.C.: U.S. Department of Health, Education, and Welfare, Food and Drug Administration).

SHERMAN, J. M. 1937. The streptococci. Bacteriol. Rev. *1*, 3.

SILLIKER, J. H., SCHMALL, A., and CHIU, J. Y. 1966. The fluorescent antibody technique as a means of detecting salmonellae in foods. J. Food Sci. *31*, 240.

SIMONSEN, B. 1966. Personal communication (Danish Meat Products Laboratory, Howitzvej 13, Copenhagen, Denmark).

SLANETZ, L. W., CHICHESTER, C. O., GAUFIN, A. R., and ORDAL, Z. J. (eds.). 1963. Microbiological quality of foods, Proc. of Conf., Franconia, N.H., 27–29 Aug. 1962 (New York: Academic Press).

SMITH, M. H. D., NEWELL, K. W., and SULIONTI, J. 1965. Epidemiology of enteropathogenic *Escherichia coli* infections in non-hospitalized children. *In* Anti-microbiological agents and chemotherapy (American Society for Microbiology, Waverly Press Inc.; Baltimore, Md.: The Williams and Wilkins Co.), pp. 77–83.

SUTTON, R. G. A. and HOBBS, B. C. 1968. Food poisoning caused by heat-sensitive *Clostridium welchii*. J. Hyg., Camb. (in press).

TAKIKAWA, I. 1958. Studies on pathogenic halophilic bacteria. Yokohama Med. Bull. *9*, 313.

TAYLOR, B. C. and NAKAMURA, M. 1964. Survival of shigellae in food. J. Hyg., Camb. *62*, 303.

TAYLOR, C. E. D. and COETZEE, E. F. C. 1966. Range of heat resistance of *Clostridium welchii* associated with suspected food-poisoning. Mon. Bull. Minist. Health *25*, 142.

TAYLOR, J. and CHARTER, R. E. 1952. The isolation of serological types of *Bact. coli* in two residential nurseries and their relation to infantile gastro-enteritis. J. Pathol. Bacteriol. *64*, 715.

TAYLOR, W. I. 1965. Isolation of shigellae. I. Xylose lysine agars; new media for isolation of enteric pathogens. Am. J. Clin. Pathol. *44*, 471.

THATCHER, F. S. 1955. Microbiological standards for foods: Their function and limitations. J. Appl. Bacteriol. *18*, 449.

———— 1963. The microbiology of specific frozen foods in relation to public health: Report of an international committee. J. Appl. Bacteriol. *26*, 266.

———— 1965. Epidemiology of salmonellosis: Large population at risk. *In* Proc. Natl. Conf. on Salmonellosis, 1964, Atlanta, Georgia (U.S. Public Health Service Publ. No. 1262 [Washington, D.C.: U.S. Department of Health, Education, and Welfare]).

———— 1966. Food-borne bacterial toxins. Can. Med. Assoc. J. *94*, 582.

THATCHER, F. S., COMTOIS, R. D., ROSS, D., and ERDMAN, I. E. 1959. Staphylococci in cheese: Some public health aspects. Can. J. Public Health *50*, 497.

THATCHER, F. S. and ERDMAN, I. E. 1966. Unpublished data (Microbiology Division, Laboratories of the Food and Drug Directorate, Ottawa).

THATCHER, F. S. and ROBINSON, J. 1962. Food poisoning: An analysis of staphylococcal toxins. J. Appl. Bacteriol. *25*, 378.

THATCHER, F. S., SIMON, W., and WALTERS, C. 1956. Extraneous matter and bacteria of public health significance in cheese. Can. J. Public Health *47*, 234.

TILTON, R. C. and LITSKY, W. 1967. The characterization of fecal streptococci. An attempt to differentiate between animal and human sources of contamination. J. Milk Food Technol. *30*, 1.

UEDA, S., SASAKI, S., and KAHUTO, M. 1959. The detection of *Escherichia coli* O-55 in an outbreak of food poisoning. Nippon Saikingaku Zasshi *14*, 48.

U.S. DEPARTMENT OF HEALTH, EDUCATION, AND WELFARE. 1965. Proc. Natl. Conf. on Salmonellosis, 1964, Atlanta, Georgia (U.S. Public Health Service Publ. No. 1262 [Washington, D.C.: U.S. Department of Health, Education, and Welfare]).

VERNON, E. 1965. Food poisoning in England and Wales. Mon. Bull. Minist. Health and Public Health Lab. Serv. *25*, 194.

VAN OYE, E. (ed.). 1964. The world problem of salmonellosis (Monographiae Biologicae, Vol. 13 [The Hague, The Netherlands: Dr. W. Junk Publishers]).

VAN SCHOTHORST, M., MOSSEL, D. A. A., KAMPELMACHER, E. H., and DRION, E. F. 1966. The estimation of the hygienic quality of feed components using an Enterobacteriaceae enrichment test. Zentr. Veterinaermed. *13B*, 273.

VON BUCHLI, K., VAN SCHOTHORST, M., and KAMPELMACHER, E. H. 1966. Untersuchungen über die hygienische Beschaffenheit von mit Wasser resp. Luft gekühltem Schlachtgeflügel. Arch. Lebensmittelhyg. *17*, 97.

WAGENAAR, R. O. and DACK, G. M. 1955. Studies of canned cheese spread experimentally inoculated with spores of *Clostridium botulinum*. Food Res. *20*, 144.

―――― 1958. Factors influencing growth and toxin production in cheese inoculated with spores of *Clostridium botulinum* types A & B. I. Studies with surface-ripened cheese type I. J. Dairy Sci. *41*, 1182.

WILLIAMS, R. E. O., LIDWELL, O. M., and HIRCH, A. 1956. The bacterial flora of the air of occupied rooms. J. Hyg. *54*, 512.

WILSON, G. S. and MILES, A. A. 1964. Topley and Wilson's Principles of bacteriology and immunity, Vol. II (5th ed.; London: Edward Arnold), p. 2509.

ZEN-YOJI, H., SAKAI, S., TERAYAMA, T., KUDO, Y., ITO, T., BENOKI, M., and NAGASAKI, M. 1965. Epidemiology, enteropathogenicity, and classification of *Vibrio parahaemolyticus*. J. Infect. Diseases *115*, 436.

PART II

A SELECTION OF
METHODS FOR THE
MICROBIOLOGICAL
EXAMINATION
OF FOODS

PRINCIPLES AND
PROBLEMS OF
SELECTION

The Committee was aware that apart from genuine scientific differences of opinion, many other difficulties can frustrate efforts to obtain agreement upon a particular method for the microbiological examination of foods for general adoption. Bias may be introduced by local familiarity with a method, by national or the originator's prejudice, by agency loyalties, or by lack of comparative information. Such bias may be difficult to dispel because of the inconvenience and cost of retraining technicians in the use of a different method. A well-trained technician with experience in interpreting the many minor differences of behaviour of microorganisms on specific media may often obtain better results from a relatively simple, familiar procedure than would a less competent technician using a more elaborate, but potentially more sensitive procedure. Further, methods possessing similar precision may differ in cost or in availability of the specific media or associated equipment required. In such instances, freedom to make local judgments on alternative procedures should not be hampered by overly rigid restrictions.

Nevertheless, there is an urgent need for agreement on what are the most satisfactory available methods for the microbiological analysis of foods. Development of international agreement on methods has been so slow, and the need for prompt action is now so evident, that the Committee concluded that the ultimate selection of a preferred method or methods for a specific purpose could best be accomplished in two stages, viz., (i) by discussion, debate, and compromise among an international group of experts; (ii) where the grounds for disagreement have an obviously scientific basis, or where inadequate data are available to form judgments, by initiation of an international programme for comparative tests of certain tentatively chosen methods. (A recent international comparison of methods used in 15 different laboratories for the determination of *Salmonella* in egg meringue powder, as sponsored by the Committee, demonstrated the great amount of planning and testing that is needed before agreement based on such experiments can be attained.)

Since provisional standard methods are needed now, and the anticipated refinements suggested by comparative testing can be introduced in due course, the procedure followed by the Committee was: (i) to review the progress in development of standard methods used in each of the 11 countries represented by Committee members; (ii) to weigh the activities of other committees with comparable objectives, such as the Microbiology Committee for the Benelux Countries, the Scandinavian Committee for Standardization of Foods, the International Standards Organization, the Food Hygiene Committee of the Codex Alimentarius Commission, the International Institute of Refrigeration, the Association of Official Analytical Chemists, specific subcommittees of the American Public Health Association, the Association of Food and Drug Officials of the United States, the British Food Manufacturing Industries Research Association, and specific industrial committees in the United States, Canada, and the United Kingdom; (iii) to appraise methods currently preferred in the laboratories of the participants; (iv) to prepare statements on useful procedures acceptable to the whole Committee.

In many instances, the present status of method development did not warrant recommendation of a single preferred method. For example, various modifications of a method for *Salmonella* are offered, dependent on the type of food to be investigated. A choice of methods is offered when several appeared to be of closely comparable value and had been regionally adopted by competent laboratories.

The Committee also emphasizes that no matter how sensitive a method may be, the validity of the conclusions to be deduced from analytical data derived from a few specimens taken from a large consignment will be dependent, in large part, upon the adequacy of the sampling plan used in conjunction with the test. This is of great importance when analysing foods in international commerce. Among the nations of the world, a particular type of food will be produced by many different manufacturers with diverse differences in production facilities, sanitation, worker hygiene, and control testing schemes. Production would be subject to different standards of supervision. An evident problem is how to devise tests that will suit the purpose of the official analysts of many countries and which, for a specific product, will have comparable meaning no matter what its origin. Such tests would need to specify both analytical procedure and a sampling plan.

The Committee intends to undertake a more detailed study of sampling related to microbiological analyses, but some important principles to be applied in sampling are outlined in the next chapter.

Limitations of methods

All laboratory methods have their limitations and are to some extent tentative. Since any given method can represent no more than the best attainable in the light of prevailing knowledge, it should be regarded as subject to replacement or modification in due course. In other words, the Committee's recommendations are of an interim nature.

A program of multilaboratory, international, comparative tests of many methods applicable to the microbiological and toxinogenic analysis of foods is being launched by the Committee. Financial support for this project, from private industry as well as from a national and an international government agency (all specified in Appendix III), is hereby gratefully acknowledged.

SAMPLING PROCEDURES

The purpose of sampling plans

Brief notes on sampling are offered in order that a few principles may be appreciated and their importance recognized in the use of the proposed analytical methods.

An adequate sampling procedure is essential whenever the investigator wishes to make deductions about a large quantity of food based on results obtained from the smaller sample examined: for example, whenever a decision must be made to accept or to reject a shipment of food on the basis of analytical results from a few specimens (Thomas and Cheftel, 1955). This need is accentuated when a method seeks to determine the presence of pathogens that may be sparsely distributed within the food, or if disposal of a food shipment will depend on the demonstrated bacterial content in relation to a legal standard.

The analytical laboratory can seldom cope with the number of samples needed to demonstrate the absence of harmful microorganisms in a lot of commercial size, at a probability level which most scientists would consider desirable. Practical considerations require that a lower degree of probability will have to be accepted in many instances. The need for compromise, however, does not justify unnecessary lack of observance of sampling principles, a fault which is common.

Choice of a sampling plan

Two general types of sampling schemes or plans are available: (1) attributes plans and (2) variables plans (Duncan, 1965; AFDOUS, 1966).

1 Attributes plans are designed to establish the proportion of units sampled which fail to meet one fixed standard. In the present context, substandard units would be those containing more than a specified number of organisms of a particular kind (see Appendix V).

2 Variables sampling plans make use of information concerning the

distribution, within a comparable class of foods, of the factor being sought, in this case, the test organism. Such information, determined experimentally for a given class of foods, is prerequisite. Some problems encountered in acquiring this information are well illustrated by Shewan and Baines (1963).

Of the two types of sampling plans, with the same analytical effort, variables plans usually provide higher probability levels than attributes plans or, alternatively, will permit a statement at a given probability level with fewer samples. However, as just noted above, a variables plan requires some knowledge of the distribution of the test organisms within the product. International diversity among food producers makes it unlikely that any common distribution can be assumed, which suggests that, for international purposes, attributes plans could be more easily adopted than variables plans.

Attributes plans as normally used distinguish between two classes of food, which, by definition, are separable into "acceptable" and "non-acceptable." Such two-class attributes plans have been applied to a number of types of foods (AFDOUS, 1966).

Dr. D. F. Bray[1] (personal communication, 1967) has proposed for consideration a possible two-level scheme based on three classes instead of the usual two for attributes plans. The three-class plan could be used to specify a number (a microbial population) which would indicate a severely unsatisfactory condition and which, if found in any specimen, would require rejection of the lot. The second number would represent a value which could be tolerated occasionally at some chosen level, say 3 out of 10 specimens, but which, if found more frequently, would be indicative of generally high counts and would be cause for rejection of the lot. The plan should be used only for organisms that are not of direct hazard. It should not be used when the standards, or classes, relate to infectious pathogens or to toxinogenic organisms such as *Clostridium botulinum*.

As an example, suppose 10 specimens are tested and it is agreed that 10^7 (per gram) would be an intolerable number to find in any specimen. If any of the 10 specimens reveals 10^7 or a greater number of organisms, then the whole lot should be condemned. If all specimens tested contain fewer than, say, 10^5, then the entire lot is acceptable. Further, suppose that 3 specimens will be tolerated within the range between the wholly acceptable and the completely intolerable levels (in this illustration from 10^5 to 10^7). This tolerated number, that is 3 of the 10, would be based

1/Chief, Statistical Services Division, Food and Drug Directorate, Ottawa, Canada.

upon criteria such as the seriousness of the contamination and the feasibility of attainment under good commercial practice. If more than 3 are found in this range, the lot would be rejected.

To develop an attributes plan for microbiological sampling, the statistician needs to know (1) the maximum number of substandard samples that would be tolerated in the production lot, and (2) the probability level to be applied to this statement. With these values, he can specify the number of specimens that must be tested from within a given lot.

To illustrate, suppose it is agreed that detection, by the method proposed, of a single pathogenic organism (e.g., *Salmonella*) will be cause for condemnation of the "lot." To the statistician this implies that the maximum number of substandard units allowed in the sample will be zero (i.e., $c = 0$).[2] Authorities could then specify acceptable probability levels (e.g., $\alpha = 0.05$, $\beta = 0.10$). Given these two pieces of information (the maximum number of defective units and probability levels), attributes sampling plans can be written which will specify the number of items that must be inspected. (Adjustments of the sample size may be required if the lot is not large, say 100 containers.) Similar illustrations could be given for different criteria, such as a plate-count standard or an MPN (most probable number) standard for coliforms, where an appropriate level would be required for each, and c might or might not be chosen equal to zero.

Alternatively, if a statement can be made as to the number of specimens it would be feasible to analyse, the statistician can calculate for the agreed number of specimens the probability of accepting a lot that contains numbers of organisms in excess of the standard, even though the values found by test may not exceed the standard. Illustrations of such calculations are given in a bulletin prepared by the Advisory Committee on the Microbiology of Frozen Foods (AFDOUS, 1966); see also Appendix V.

The probability attributes of a sampling plan can be expressed by an "operating characteristic curve" (O.C. curve[3]), characteristic for a particular type of product, and from which the probability of accepting a lot of given quality can be read. A number of such curves are illustrated in the same bulletin (AFDOUS, 1966, pp. 69–71). An O.C. curve for *Salmonella* in egg products is shown in a report of the Subcommittee on Sampling and Methodology (1966); see also Appendix V.

2/c is the number of defective units which cannot be exceeded without the lot being rejected, α is the probability of declaring that the lot is not acceptable when it really is, and β is the probability of declaring that the lot is acceptable when it really is not.
3/The operating characteristic curve shows the relation between the probability of accepting the lot and the actual proportion of defective units in the lot.

The data and calculations presented by Shelton *et al.* (1962) suggest that for several frozen foods little additional precision is obtained by increasing the number of specimens (containers, etc.) from a lot to more than 10, but analysis of 10 specimens provides a much more dependable result than if only 6 are used. Use of 10 specimens was recommended for the type of sampling plan employed in that study. Van Schothorst *et al.* (1966) describe a modification in which the examination of 10 initial samples from a lot may lead to (i) acceptance, (ii) rejection, or (iii) examination of an additional 10 samples before coming to a final decision.

A number of "rule-of-thumb" sampling plans are in use, such as the testing of a number of specimens equivalent to the square root of the number of containers in the "lot." Such plans may sometimes indicate an appropriate number of samples, but frequently will not.

Sampling of heterogeneous foods

The degree of homogeneity of the food being tested greatly influences the choice of sampling plan, more samples being needed if the food is not homogeneous (Subcommittee on Sampling, 1966).

Large containers present special sampling problems because microorganisms are frequently unevenly distributed within the container. If products are in bulk, or in containers of a size making it impossible to transport them to the laboratory, random samples should be taken from different parts of the container. The *number* of subsamples should be decided according to statistical (mathematical) considerations.

Specific instructions for drawing samples should be provided to take account of the nature of the product and the nature of the container.

"Stratification" and homogeneity

If, in a lot that has been submitted for examination, there are large containers from which samples must be taken, the analyst commonly takes an equal number of samples (e.g., core samples from drums of powder, small packages from large packages) from each container. The only reason for doing this is that more variability is to be expected from one large container to another than from one sample to another within a container. "Stratification," then, is a device for partitioning variation so as to put as much variation as possible between strata, leaving the "within-strata" samples relatively homogeneous. Recommendations on methods of drawing samples with specific instructions regarding stratification could be written for various products, depending on the nature of the containers and the anticipated degree of heterogeneity.

Sensitivity of the methods

The classification of a sample as satisfactory or not, with respect to the specified number of organisms to be tolerated in it, depends partly on the sensitivity of the method used for examination. The method chosen must obviously be able to classify the product accurately relative to the tolerance specified. For example, if we choose an "action" level of 10 organisms per gram, the method must be capable of detecting as few as 10. On the other hand, if we choose an action level of 10 and have at our disposal two methods, one capable of detecting 5 and another 1 organism per gram, either would serve and choice would depend on other factors, such as cost. If the presence of a pathogen in numbers fewer than a method-determined limit is thought to have public health significance, then a more sensitive method should be sought, or larger test aliquots should be examined.

If the sensitivity of the method used is known, the limit of tolerance can be set at the limit of determination by the method or at some arbitrary value above this. But, for a given food, the sensitivity of a method is rarely known precisely, because changes in the nature of the food or in its microbial ecology can modify it, sometimes severalfold. Nevertheless, experience usually indicates the range of sensitivity for a method applied to particular types of food. To adjust to this uncertainty, it is often practicable to apply a limit of acceptability in the form of an expression such as "failure to find" a given pathogen "as determined by the standard method," when employing a defined sampling plan.

Selection of criteria

Ultimately, a judgment must be made as to what constitutes a limit that has meaning, in relation to the intention of the control being exercised. The specification of the limit can modify both the choice of sampling plan and the sensitivity required in the analytical method.

For non-pathogens, a limit is usually based on data derived from surveys which show what values can be attained under acceptable commercial conditions. Whether or not a precise figure or a range is specified is essentially a matter of arbitrary judgment by the controlling authority.

For bacteria causing non-infectious food poisoning which may depend on toxin production (e.g., *Staphylococcus aureus* and *Clostridium botulinum*), or on some other mechanism dependent on large numbers of organisms (e.g., *Clostridium perfringens* and *Bacillus cereus*), the numbers should be limited. In addition, the level must be so low that during handling or processing it is unlikely that multiplication will give rise either to significant amounts of toxin or to a dose of organisms able to initiate symptoms of food poisoning. Thus, for example, a limit recommended by the Committee (Thatcher, 1963) was 100 viable *Staph. aureus* bacteria per gram in frozen cooked seafood.

For infectious pathogens (e.g., salmonellae and shigellae), none should be present; but such an ideal situation is impossible to determine, and hence such a statement or requirement has little practical meaning. Therefore the application of a standard should be based on a reasonably sensitive sampling plan and method of examination.

The application of such requirements has led to a substantial improvement in the microbiological quality of products and to a marked reduction in the distribution of pathogens.

If the intention is to determine the adequacy of sanitation practised during processing of a food, the limits prescribed should be related to the numbers of organisms present after the processing. Thus, a time element is involved. *Escherichia coli* and staphylococci may die rapidly in certain types of cheeses. In other foods, sliced meats for instance, these organisms may proliferate if the ambient temperatures permit. Hence, the interpretation of microbiological findings may need consideration of such additional factors as temperatures experienced by the foods and the likelihood that either an increase or a decrease in numbers of particular organisms may have occurred during the time interval since processing. If the over-all hygienic quality of a food is being assessed, the presence of large numbers of organisms due to growth after processing represents a hygienic fault. On the other hand, indicator or toxinogenic organisms may die during storage, and the low numbers found on testing could give an erroneous impression of safety, although, if there had been very poor sanitation, toxins might remain—for example, from staphylococci after the death of the organisms.

Specificity of sampling procedures in relation to food type

The ultimate deduction is that, in practice, each type of food requires a particular sampling procedure, in which the number of samples to be tested depends on the normal variation of microbial contamination, the homogeneity of the food, and the sensitivities of the methods of examination relative to the standards to be applied. It follows that, to obtain distribution data suitable for use in the design of sampling plans for the many different types of food that need analysis, it is necessary to undertake analytical surveys using truly random samples. Careful design of such surveys is essential, the system to be adopted depending on the variables to be analysed.

RECOMMENDED METHODS FOR MICROBIOLOGICAL EXAMINATION

The Committee wishes to emphasize that the methods recommended here are proposed as interim methods until comparative interlaboratory testing can be carried out. The methods selected are those which are known to have been widely used, and which within the experience of the Committee have been the more generally acceptable. The proposals are tentative and, as such, are subject to cancellation or modification as new research and comparative studies may dictate. The Committee has initiated a scheme of international interlaboratory testing to establish with greater precision than is now feasible the relative merits of the alternative methods described.

The methods are recommended primarily for use in tests intended to provide the basis for legal judgments on the microbiological quality and safety of food, particularly of foods moving in international commerce. In making selections, therefore, emphasis has been placed on sensitivity and reproducibility. It is fully recognized that for many local uses such as routine surveillance and factory control and for "screening" purposes, for some of the categories considered methods exist that require fewer media and manipulations, provide results more rapidly, and are cheaper than those proposed. These, however, may lack the precision desirable under arbitrational circumstances. The Committee intends to deal with such methods in a subsequent publication.

For several categories (preparation and dilution of the food homogenate, and enumeration of mesophilic aerobes, coliforms, staphylococci, and *Clostridium perfringens*) the Committee has recommended more than one method. In each case the alternative methods are judged to be equally suitable for the purposes intended, but may represent strong regional preferences often based on satisfactory usage by large numbers of technicians. No significance is attached to the order in which the methods are listed.

It is assumed that the methods will be used by, or the users directed by, a competent microbiologist. The procedures applied to the preparation, organization, sterilization, aseptic technique, controlled testing, and the proper cleaning and use of equipment and materials, therefore, must be commensurate with good laboratory practice. It is assumed, also, that the quality of equipment and materials used to carry out the various procedural steps will be such as to permit the degree of accuracy specified. The descriptions of the methods are, therefore, as free from non-essential detail as is practical for a working manual.

For detailed specifications on apparatus and materials, the reader is referred to *Standard Methods for the Examination of Water and Waste-water* (APHA *et al.*, 1965) and to *The Bacteriological Examination of Water Supplies* (Ministry of Health and Ministry of Housing and Local Government, 1956). For recommendations on safety procedures in the laboratory, see Appendix IV.

Preparation and dilution of the food homogenate

Methods of isolation and enumeration of the microorganisms present in all non-liquid foods usually require preliminary treatment of samples of the food to release into a fluid medium those microorganisms which may be embedded within the food or within dried or gelatinous surface films. The usual practice is to use an electrically driven mixing device with cutting blades revolving at high speed. A homogeneous suspension of food and microorganisms is thus prepared which permits the preparation of dilutions appropriate for use in various determinative or enumerative procedures. Standardization of this preliminary procedure is important. Excessive speed of the cutting blades or unduly prolonged use of the mixer may cause injury to microbial cells, either mechanically or by the heat generated. Thus excessive mixing may result in a reduced "count"; insufficient mixing may not release embedded bacteria or may provide a heterogeneous distribution within the supension. Limitation of the speed of the cutting blades and of the duration of mixing time is necessary to provide optimum results. The times and speeds recommended below were determined from many years of experience and experimental determinations.

Care must be taken to minimize risk from aerosols when pathogens may be present in the test food. All mechanical blendors create aerosols.

Another factor highly important to the accurate enumeration is the nature of the diluent to be used. Many commonly used diluents, such as tap or distilled water, saline solution, phosphate buffers, and Ringer solution, have been shown to be toxic to certain microorganisms, particularly if the time of contact is unduly prolonged (Winslow and Brooke, 1927; Butterfield, 1932; de Mello *et al.*, 1951; Gunter, 1954; Stokes and Osborne, 1956; King and Hurst, 1963). The presence of food material may protect many organisms from this toxicity, but one can by no means depend on this (Hartman and Huntsberger, 1961; Hauschild *et al.*, 1967). Progressive dilution would also offset such a potential effect. A general purpose dilution medium of growing favour is 0.1% peptone (Straka and Stokes, 1957; King and Hurst, 1963). When enumerating *C. perfringens*, Hauschild *et al.* (1967) have shown a pronounced advantage in using peptone solution in comparison with standard phosphate buffer diluent (APHA, 1960).

Where reproducibility rather than maximum sensitivity is the main objective in standard procedures (as for the plate count), various diluents may be adequate provided they are used in accordance with a defined composition and period of exposure, but for the isolation and enumeration of particular species or groups known to be sensitive to diluents containing inorganic salts or to distilled water, the least toxic diluent is to be preferred. Present knowledge suggests that 0.1% peptone best meets this need. The validity of regional or personal preferences for particular diluents needs further investigation before a diluent can be defined for all enumerative purposes.

Method 1[1]

I. Apparatus and materials

1 Mechanical blendor, two-speed model or single speed with rheostat control.

2 Glass or metal blending jars of 1 litre capacity, with covers, resistant to autoclave temperatures. One sterile jar (autoclaved at 121° C for 15 minutes) for each food specimen to be analysed.

3 Balance with weights. Capacity at least 2500 g; sensitivity 0.1 g.

4 Instruments for preparing samples: knives, forks, forceps, scissors,

1/The method follows closely that described in *Standard Methods for the Examination of Dairy Products* (APHA, 1960).

spoons, spatulas, or tongue depressors, sterilized previous to use by autoclaving or by hot air.

5 A supply of 10- or 11-ml pipettes (milk dilution specifications;
APHA, 1960).

6 Phosphate-Buffered Dilution Water (Medium 62, Part III) or Peptone Dilution Fluid (Medium 53, Part III). Sterilized in autoclave for
each specimen: (i) 450 ml in flask or bottle, (ii) 90 or 99 ml blanks in
milk dilution bottles (APHA, 1960, specifications) or similar containers.

II. Procedure

1 If the sample is frozen, thaw it in its original container (or in the
container in which it was received at the laboratory) for 18 hours in a
refrigerator at 2–5° C. If the frozen sample can be easily comminuted
(e.g., ice cream), proceed without thawing.

2 Tare the empty sterile blendor jar, then weigh into it 50 ± 0.1 g
representative of the food specimen. If the contents of the package are
obviously not homogeneous, (as, for example, a frozen dinner) a 50 g
sample should be taken from a macerate of the whole dinner, or each
different food portion should be analysed separately, depending upon the
purpose of the test.

3 Add to the blendor jar 450 ml of Phosphate-Buffered Dilution
Water or Peptone Dilution Fluid. This provides a dilution of 10^{-1}.

4 Blend the food and dilute promptly. Start at low speed, then switch
to high speed within a few seconds; or gradually within a few seconds
increase to full voltage. Time the blending carefully to permit 2 minutes at
high speed.[2] Wait 2 or 3 minutes for foam to disperse. However, once the
diluent has been weighed into the sample, no unnecessary delays should
be permitted.

5 Measure 1 ml of the 10^{-1} dilution of the blended material, avoiding
foam, into a 99 ml dilution blank, or 10 ml into a 90 ml blank.
Shake this and all subsequent dilutions vigorously 25 times in a 1-foot arc.
Repeat this process using the progressively increasing dilutions to prepare
dilutions of 10^{-2}, 10^{-3}, 10^{-4}, and 10^{-5}, or those which experience indicates as desirable for the type of specimen tested. If Phosphate-Buffered
Dilution Water is used as the diluent, no more than 15 minutes should
elapse from the time the 10^{-1} dilution is prepared until the suspended food
is in the appropriate media. If Peptone Dilution Fluid is used, see step 5
of Procedure for Method 2 below.

2/For certain foaming foods, such as cream pies, a 1 minute blend is sufficient.

Method 2[3]

I. Apparatus and materials

1 Mechanical blendor, operating at not less than 8000 rpm and not more than 45,000 rpm.

2 The requirements for Method 1 above, I, items 2 through 4.

3 A supply of 1-ml bacteriological pipettes.

4 Culture tubes for dilution fluid, 15–20 ml. capacity suitable.

5 Peptone Dilution Fluid (Medium 53, Part III).

II. Procedure

1 Begin the examination as soon as possible after the sample is taken. Refrigerate the sample at 0–5° C whenever the examination cannot be started within 1 hour after sampling. If the sample is frozen, thaw it in its original container (or in the container in which it was received in the laboratory) in a refrigerator at 2–5° C and examine as soon as possible after thawing is complete or after sufficient thawing has occurred to permit suitable subsamples to be taken.

2 Weigh into a tared blendor jar at least 10 g of sample, representative of the food specimen.

3 Add 9 times as much dilution fluid as sample. This provides a dilution of 10^{-1}.

4 Operate the blendor according to its speed for sufficient time to give a total of 15,000 to 20,000 revolutions. Thus, even with the slowest blendor the duration of grinding will not exceed 2.5 minutes.

5 Allow the mixture to stand for 15 minutes at room temperature to permit resuscitation of the microorganisms.

6 Mix the contents of the jar by shaking, and pipette duplicate portions of 1 ml each into separate tubes containing 9 ml of dilution fluid. Carry out steps 7 and 8 below on each of the diluted portions.

7 Mix the liquids carefully by aspirating 10 times with a pipette.

8 Transfer with the same pipette 1.0 ml to another dilution tube containing 9 ml of dilution fluid, and mix with a fresh pipette.

9 Repeat steps 7 and 8 until the required number of dilutions are made. Each successive dilution will decrease the concentration 10-fold.

3/This method is used widely, particularly in European countries. As described, it follows closely the procedure being recommended by the International Standards Organization.

Enumeration of mesophilic aerobes: The agar plate colony count

The number of aerobic mesophilic microorganisms ("plate count") found in foods has been one of the more commonly used microbiological indicators of the quality of foods, except where direct fermentation (as in cheese and certain types of sausage) or a natural "ripening" process gives rise to very large number of bacteria. The aerobic mesophilic count has limited validity for assessing safety (cf. Part I, p. 25). Nevertheless, it has value with many foods: in indicating the adequacy of sanitation and temperature control during processing, transport, and storage; in forming an opinion about incipient spoilage, probable shelf-life, uncontrolled thawing of frozen foods, or failure to maintain refrigerative temperatures for chilled foods; and in revealing sources of contamination during manufacture. The "plate count" has special application to imported foods, where the importing country has no opportunity to control the standard of sanitation practised in the manufacturing establishments.

The three methods in common use for enumerating aerobic mesophilic microorganisms are the "Pour Plate" or "Standard Plate Count" (recently reviewed by Hartman and Huntsberger (1961) and Angelotti (1964)), the "Drop Plate" (Miles and Misra, 1938), and the "Surface Plate" or "Spread Drop" methods (Reed and Reed, 1948; Campbell and Konowalchuk, 1948; Buck and Cleverdon, 1960; Gaudy *et al.*, 1963; Clark, 1967). None of the procedures can be depended upon to enumerate all types of organisms present within the test specimens. Many cells may not grow because of specifically unfavourable conditions of nutrition, aeration, temperature, or duration of incubation.

Whenever the methods are used, the temperature must be specified, but the procedures are applicable for the examination of different groups of microorganisms having different temperature ranges; such as 0–5° C for psychrophiles, 30–35° C for mesophiles, and 55° C for thermophiles. The temperature chosen would depend upon the purpose of the examination: for example, the degree of spoilage and probable shelf-life of ice-packed chickens would best be estimated by using an incubation temperature of 0–5° C; 35° C would provide the most information about general plant sanitation; but 55° C would be the preferred incubation temperature for determining the thermophilic contamination from blanching equipment or from heated pumps in a food conduit system. Accurate control of the incubation temperature is essential in each procedure. However, because microorganisms have very diverse temperature limits for growth, no one incubation temperature will absolutely exclude all organisms from another group.

Within the mesophilic range (the one of chief concern when a food is investigated in the interests of public health), the number of colonies obtained decreases rapidly as the incubation temperature increases from 35° to 40° C. Temperature control is therefore critical in this range and, to obtain the needed precision, incubators are required with minimal fluctuation or variation of temperature throughout the incubation chamber. Some commercial incubators do not meet the requirements. All incubators should be checked and calibrated frequently.

In appraising the three methods, the Committee appreciates the saving in cost obtained by using the Surface Plate or Drop Plate procedures: less medium and fewer Petri plates are required, and less expensive pipettes are used. For this reason, many laboratories prefer to use these methods, particularly for routine analyses of a "screening" nature.

Apart from cost, the advantages and disadvantages of the Drop Plate and Surface Plate methods in comparison with the Pour Plate method are listed below.

Advantages (i) It is not essential that the medium be translucent, as it is for the Pour Plate method. (ii) Because all colonies develop on the surface of the medium, colony appearance can generally be studied more satisfactorily, and the proportion of the various colony types can be estimated more readily. (iii) By dropping several dilutions on a plate, the accuracy of the dilution system can be easily assessed. (iv) Heat-labile organisms are not killed, as may occur when cells are exposed to melted agar (45° C) in the Pour Plate method. (v) Pre-prepared plates can be stored and moved easily within a laboratory or to field laboratories for "on the site" studies.

Disadvantages (i) Because of the small volume of specimen used (0.02 or 0.1 ml), suspended food could interfere by causing a nutrient or inhibitory effect on microorganisms, and by causing optical confusion between food particles and small colonies except at the higher dilutions of the specimen. (ii) For the same reason, the Drop Plate and Surface Plate procedures are not satisfactory for specimens with few organisms (e.g., less than 500/g), the limit being 10–100 times higher than with the Pour Plate method; however, specifications and standards for the mesophilic organisms in foods are likely to be well within the "countable" range. (iii) The Surface Plate method is not widely used, though it may become so because of the preference of the International Standards Organization.

The Pour Plate or Standard Plate Count procedure is used in many countries and almost exclusively so in North America, where it has been specifically recommended for dairy products (APHA, 1960) and for frozen foods (APHA, 1966; AFDOUS, 1966); it is used in most public health and food control agencies in the United States and Canada. A modification of the method, used for some years in France and regions influenced by French technology, has been adopted by the International Dairy Federation (1958) for determination of aerobic mesophiles in liquid and dried milk. In the method, milk is added to Standard Methods Agar (Medium 77, Part III) and the poured plates are incubated at 30° ± 1° C for 3 days. The Drop Plate method is used by some laboratories in the United Kingdom. The Surface Plate method is used widely in Europe and may soon be officially recommended by the International Standards Organization for use with meat and meat products containing more than 3000 organisms per gram (Barraud *et al.*, 1967).

The standard Pour Plate procedure is so well established, and its accuracy and limitations so well understood, that the Committee recommends its retention, for purposes of arbitration. Nevertheless, one of the other methods may be chosen later, if warranted by comparison with the Pour Plate method under sufficiently diverse circumstances. The Committee will undertake international testing to compare the relative accuracy of different plating procedures.

Method 1[4] *(The Pour Plate or Standard Plate Count Method)*

I. Apparatus and materials

1 Requirements for Preparation and Dilution of the Food Homogenate, Method 1 or 2, above.

2 Petri dishes, glass (100 × 15 mm) or plastic (90 × 15 mm).

3 1-, 5-, and 10-ml bacteriological pipettes (milk dilution specifications; APHA, 1960).

4 Water bath or air incubator for tempering agar, 45° ± 1° C.

5 Incubator, 35° ± 1° C. (*Note* The temperature constancy and uniformity of air incubators are notoriously bad. To check constancy, continuous or frequent interior temperature readings should be made over a period of several hours by suitable thermocouples or thermometers. To check uniformity, similar readings over a period of several hours should be made at various positions in the incubator while it is filled with agar plates (see also APHA, 1960).)

4/The method follows that recommended for dairy products (APHA, 1960) and for frozen foods (Lewis and Angelotti, 1964; AFDOUS, 1966).

6 Colony counter (Quebec dark field model or equivalent recommended).

7 Tally register.

8 Standard Methods Agar (Plate Count Agar: Medium 77, Part III).

II. Procedure

1 Prepare food sample by one of the two procedures recommended in the section on Preparation and Dilution of the Food Homogenate (p. 60).

2 To duplicate sets of Petri dishes, pipette[5] 1 ml aliquots from 10^{-1}, 10^{-2}, 10^{-3}, 10^{-4}, and 10^{-5} dilutions, and a 0.1 ml aliquot from the 10^{-5} dilution to give 10^{-1} to 10^{-6} g of food per Petri dish. The above series is suggested in cases where the approximate range of bacterial numbers in the specimen is not known by experience. The level of dilution prepared and plated can be varied to fit the expected bacterial count.

3 Promptly pour into the Petri dishes 10–15 ml of Standard Methods Agar, melted and tempered to $45° \pm 1°$ C.

4 Immediately mix aliquots with the agar medium by tilting and rotating the Petri dishes. A satisfactory sequence of steps is as follows: (*a*) tilt dish to and fro 5 times in one direction, (*b*) rotate it clockwise 5 times, (*c*) tilt it to and fro again 5 times in a direction at right angles to that used the first time, and (*d*) rotate it counter-clockwise 5 times.

5 As a sterility check, pour one or two plates with agar medium alone (control(s)).

6 After the agar has solidified, invert the Petri dishes and incubate them at $35° \pm 1°$ C for 48 ± 3 hours.

7 Using the colony counter and tally register, count all colonies on plates containing 30–300 colonies.

8 Compute the number of mesophilic aerobes per gram of specimen.

Method 2[6] (The Surface Plate Method)

I. Apparatus and materials

1 The requirements for Method 1 above, I, items 1, 2, 4, 6, 7, and 8.

2 A drying cabinet or incubator for drying the surface of agar plates, preferably at 50° C (see step 9, p. 158, for note on drying plates).

5/Pipettes must not be used to deliver less than 10% of their volume.

6/This method is widely used, particularly in Europe, and as described, follows closely that being recommended by the International Standards Organization (Barraud *et al.*, 1967). The method is not recommended for samples in which the number of mesophilic organisms is likely to be lower than 3000 per g.

3 Incubator, $30° \pm 1°$ C.

4 Glass spreaders (hockey-stick-shaped glass rods); approximate measurements: 3.5 mm diameter, 20 cm long, bent at right angles 3 cm from one end.

5 Graduated bacteriological pipettes, 1 ml capacity, with subdivisions of 0.1 ml or less.

II. Procedure

1 Add 15 ml of melted, cooled (45–60° C) Standard Methods Agar to each Petri dish used and allow to solidify. Dry agar plates preferably at 50° C for 1.5–2 hours. If prepared in advance the plates should not be kept longer than 24 hours at room temperature or 7 days in a refrigerator at 2–5° C.

2 Prepare food samples by one of the two procedures recommended in the section on Preparation and Dilution of the Food Homogenate (p. 59).

3 Using only one pipette, transfer 0.1 ml of each of the dilutions tested (test at least three, even if the approximate range of numbers of organisms in the food specimen is known) to the agar surface of each of two plates. Start with the highest dilution and proceed to the lowest, filling and emptying the pipette three times before transferring the 0.1 ml portion to the plate.

4 Spread the 0.1 ml portions as quickly as possible carefully on the surface of the agar plates using glass spreaders (use a separate spreader for each plate). Allow the surfaces of the plates to dry for 15 minutes.

5 Incubate the plates inverted for 3 days at $30° \pm 1°$ C.

6 Count all colonies on plates containing 30–300 colonies. If available, use colony counter and tally register for convenience.

7 Compute the number of mesophilic aerobes per gram of specimen.

Method 3[7] (The Drop Plate Method)

I. Apparatus and materials

1 The requirements for Method 1 above, I, items 1, 2, 3, 5, and 8.

2 A drying cabinet or incubator for drying the surface of agar plates, preferably at 50° C.

7/The method is a modification of the technique described by Miles and Misra (1938). It is used in a number of laboratories, among them The Food Hygiene Laboratory, Central Public Health Laboratory, Colindale, London, England. It is not recommended for food specimens in which the number of mesophilic organisms is likely to be lower than 3000 per g.

3   Pipettes calibrated to deliver 50 drops per ml of distilled water at 20° C at the rate of 1 drop per second when held vertically.[8] Each drop delivers 0.02 ml.

4   Horse-Blood Agar (Medium 30, Part III).

5   Ringer Solution, 1/4 strength (Medium 65, Part III).

II. Procedure

1   Prepare dried plates of Horse-Blood ·Agar or Standard Methods Agar as described in II, step 1, of Method 2 above.

2   Mark the plates on the bottom into three equal segments, and indicate for each segment the dilution to be used.

3   Prepare food samples by one of the procedures recommended in the section on Preparation and Dilution of the Food Homogenate (p. 60), using 1/4 strength Ringer Solution as the diluent.

4   Using the specially calibrated pipette, deliver 2 separate drops (0.02 ml each) of each dilution on the surface of the relevant segment. Thus, with only two plates, six dilutions ($10^{-1}, 10^{-2}, \ldots, 10^{-6}$) can be plated in duplicate.

5   Allow the drops to dry for about 15 minutes at room temperature with the agar surface uppermost.

6   Incubate the plates inverted for 48 hours at $30° \pm 1°$ C.

7   Count all colonies for dilutions giving 20 colonies or less per drop.

8   Compute the number of mesophilic organisms per gram of specimen.

Coliform bacteria

Coliform organisms, "faecal coliforms," and *Escherichia coli* are primarily used to indicate some degree of potentially hazardous contamination, based on the assumption that the natural habitat of the family of bacteria to which they belong (the Enterobacteriaceae) is the faeces of man and other mammals. *E. coli* is generally accepted as the most positive indicator

8/Pipettes which deliver 0.02 ml per drop can be made easily in the laboratory as follows. Heat evenly the middle 1.5 cm of a 12–16 cm length of 5 mm OD standard wall glass tubing and draw out smoothly to form a fine, tapered capillary. Cool, cut in two at the middle, and insert the tapered end of each piece, in turn, in the hole of a 0.95-mm-diameter wire gauge (available from R. B. Turner and Co., Inocula House, Church Lane, London N. 2) as far as it will go. Cut capillary off, flush with gauge, on the side nearest the pipette shoulder. The capillary part of the pipette should be straight and spherical with the cut end at right angles to the sides and having no protruding ridges or chips.

of faecal contamination; a population of faecal coliforms is likely to contain a high proportion of *E. coli* or variants probably of recent faecal origin. Coliforms include, in addition to *E. coli*, the various species of *Enterobacter* (*Aerobacter*) and *Klebsiella*, some of which may grow or persist for long periods in non-faecal habitats. They are best used to indicate post-processing contamination and some degree of inadequate sanitation in those foods which have received a process sufficient to have removed such organisms.

The faecal connotation is important. Contamination of a food with *E. coli* implies a risk that one or more of a wide diversity of enteric pathogens may also have gained access to the food and hence introduced a health hazard. Poor factory sanitation, including inadequate worker hygiene, is commonly involved in such contamination.

The three groups (coliform organisms, faecal coliforms, and *E. coli*) represent the organisms that are recovered by the use of three successive refinements of determinative methodology, each progressively more complex and costly. With relatively few exceptions, such as the specific serotypes of *E. coli* which cause epidemic infantile diarrhoea, these organisms are not pathogens. Their indicator role is the important one.

Three major methods are in use for the determination of coliforms. One is based on an MPN (most probable number) procedure using Lauryl Sulphate Tryptose Broth (LST) followed by confirmation of gas-positive tubes using Brilliant-Green Lactose Bile Broth (BGLB), each being incubated at 35° C. The second method uses MacConkey Broth, alone, incubated at 37° C. The third method uses BGLB at 35–37° C, followed by confirmation on Violet Red Bile Agar or Endo Agar. The three methods are described here because each has been used successfully for its purpose by many laboratories. Further, such comparative data as exist on the relative merits of each tend to be conflicting. Definitive data were insufficient to establish to what extent the various methods were measuring the same organisms or recording the diverse types in comparable proportion. The Committee plans to make such comparative determinations.

LST broth was chosen in preference to Lactose Broth chiefly because of the results obtained from an extensive analysis of many types of precooked frozen foods, in which LST, in conjunction with BGLB, yielded 36% more positive results than Lactose Broth (Shelton *et al.*, 1962). Many reports of similar advantages occur in the literature as reviewed by the Advisory Committee on the Microbiology of Frozen Foods (AFDOUS, 1966) and by Hall (1964).

Faecal coliforms ("faecal coli") represent a population presumed to contain a high proportion *E. coli*, but without the actual proportion of

E. coli type I being positively established. The purpose of the test is to indicate that the organisms recorded by the methods employed were derived from the faeces of warm-blooded animals.

Two methods for determining faecal coliforms have been selected. Recommendation of a single method was not feasible, again owing to lack of comparative data.

One method uses E. C. Broth and a carefully controlled temperature of 45.5° ± 0.2° C. The inoculum is derived from gas-positive tubes of LST Broth. Other elevated temperatures have been used (44° and 44.5° C), but comparative studies by Fishbein (1961), Fishbein and Surkiewicz (1964), and Fishbein *et al.* (1967) establish a strong preference for 45.5° C. The last authors reported that the respective rates of recovery of *E. coli* types I and II from coliforms derived from E.C. Broth incubated at 45.5° and 44.5° C were 77.2% and 53.1%, even though 4% of the *E. coli* cultures that grew at 44.5° C failed to grow at 45.5° C. Three times as many false positives were recorded at 44.5° C than at 45.5° C.

Recently, Geldreich (1966) has published a comparison of nearly 12,000 cultures of coliforms, isolated from faeces and polluted soils or from soils not known to be polluted. Geldreich recommended the elevated incubation temperature of 44.5° (±0.5°) C as offering the best compromise between acceptable sensitivity and specificity for the faecal types. The Committee believes that the apparent conflict between this conclusion and the preference for 45.5° C cited above, which was based on cultures derived from foods, can only be resolved by further experiment.

A second method, widely used in continental Europe (Buttiaux *et al.*, 1956; Guinée and Mossel, 1963), employs BGLB incubated at 44° ± 0.1° C and Peptone Water, the inoculum being derived from positive tubes of MacConkey Broth. This is the method of Mackenzie *et al.* (1948). The usual means for confirmation of *E. coli* is by use of the IMViC tests as described below.

Most strains of *E. coli* are not pathogenic to man. The exception is a group of serologically identifiable strains which have been known for years to cause infantile diarrhoea (Neter *et al.*, 1951; Taylor and Charter, 1952; Ewing *et al.*, 1957, 1963). The epidemic form of the illness is particularly well known in nurseries and hospitals. More recently, some of these same strains have been identified from diverse foods and in a few instances have been associated with well-defined cases of food-poisoning (Costin *et al.*, 1960, 1964; Koretskaia and Kovalevskaia, 1958). Because of this, it appears desirable to determine their presence in food. Since the enteropathogenic *E. coli* (EEC) are physiologically typical *E. coli*, any method designed to enumerate or detect coliform organisms or faecal coliform organisms will include the EEC. With some foods, pre-enrichment or resuscitation may be helpful. For this purpose the food may be incubated in Peptone Water for 2 hours at 37° C, and an equal volume of double-strength MacConkey Broth then added before incubation at 44° C. Acid and gas production indicates the presence of *E. coli*, and the culture is streaked on to Eosin Methylene-Blue Agar or MacConkey Agar. Alternatively, it is suggested that gas-positive Lauryl Sulphate Tryptose, E.C., Brilliant-Green Lactose Bile 2%, or MacConkey broth tubes from food samples be streaked to Eosin Methylene-Blue Agar or MacConkey Agar plates.

A suitable number (up to 10, if present) of typical *E. coli* colonies from these plates must then be tested with O and OB antisera to determine the presumptive presence of EEC. Complete serological identification requires further testing with H antisera. Methods for the preparation of reagents, production of antisera, and performing the necessary tests have been described by Edwards and Ewing (1962). Because of the interest of physicians and medical microbiologists in the EEC, antisera are available for research purposes from commercial and/or government sources in several parts of the world. Where such reagents are accessible, the food microbiologist should have little technical difficulty in applying these materials and methods to isolates obtained from foods. However, uniform procedures cannot be recommended for international use until there is better agreement regarding the extent of serological testing that is necessary, and until the required antisera are more generally available to control laboratories.

ENUMERATION OF COLIFORMS:
DETERMINATION OF THE MOST PROBABLE NUMBER
(MPN)

Method 1[9]

I. Apparatus and materials

1   Requirements for Preparation and Dilution of the Food Homogenate, Method 1 or 2 (p. 60).

2   Incubator, 35° ± 1° C.

3   1-ml bacteriological pipettes.

4   Inoculating needle, with nichrome or platinum-iridium wire.

5   Brilliant-Green Lactose Bile Broth 2% (Medium 10, Part III), 10

9/The method follows that recommended for water (APHA *et al.*, 1965) and foods (APHA, 1966; Lewis and Angelotti, 1964; AFDOUS, 1966; USFDA, 1966).

ml volumes in 150 × 15 mm tubes containing inverted Durham fermentation tubes (75 × 10 mm).

6   Lauryl Sulphate Tryptose Broth (LST Broth; Medium 36, Part III), 10 ml volumes in 150 × 15 mm tubes containing inverted Durham fermentation tubes or similar glass vials (75 × 10 mm).

7   Eosin Methylene-Blue Agar (Medium 23, Part III) or Endo Agar (Medium 21, Part III), for plates.

II. Procedure

1   Prepare food samples by one of the procedures recommended in the section on Preparation and Dilution of the Food Homogenate (p. 60). All techniques of dilution should be the same. Materials remaining from dilution blanks employed in the determination of the plate count can be used.

2   Pipette 1 ml of each of the decimal dilutions of food homogenate to each of the three separate tubes of LST Broth.

3   Incubate tubes at 35° ± 1° C for 24 and 48 hours.

4   After 24 hours, record tubes showing gas production.[10] Return tubes not displaying gas to incubator for an additional 24 hours.

5   After 48 hours, record tubes showing gas production.[10]

6   Select the highest dilution in which all three tubes are positive for gas production and the next two higher dilutions. If this is not possible, because none of the dilutions yielded three positive tubes or because further dilutions were not made beyond the one yielding three positive tubes, select the last three dilutions and record the number of positive tubes in each dilution. For example, if the last three positive dilutions were 1:100, 1:1000, and 1:10,000 and the number of positive tubes in each dilution were 3, 1, and 0, respectively, the results are recorded as 1:100 dilution = 3, 1:1000 dilution = 1, and 1:10,000 dilution = 0.

7   Confirm that the tubes of LST Broth selected in step 6 above are positive for coliform organisms by transferring a loopful of each to separate tubes of Brilliant-Green Lactose Bile Broth 2% or by streaking Eosin Methylene-Blue or Endo agar plates. Incubate confirmatory tubes 24 and 48 hours at 35° ± 1° C and note gas production. The formation of gas confirms the presence of coliform organisms. Observe solid confirmatory media for typical coliform colonies after 24 and 48 hours at 35° ±

10/If tests for faecal coliforms are to be made, select a number of gas-positive tubes and follow the procedure described in Method 1 of the section on Determination of Coliform Organisms of Faecal Origin below.

1° C. The formation of black or black-centred colonies on Eosin Methylene-Blue Agar with transparent colourless peripheries or the formation of red colonies surrounded by red haloes on Endo Agar confirms the presence of coliform organisms.

8 Record the number of tubes in each dilution that were confirmed as positive for coliform organisms.

9 To obtain the MPN, proceed as follows. Determine, from each of the three selected dilutions, the number of tubes that provided a confirmed coliform result. Refer to the MPN table (Table 2) and note the MPN

TABLE 2*

MPN index and 95% confidence limits for various combinations of positive and negative results when three 10-ml portions, three 1-ml portions, and three 0.1-ml portions are used

Number of tubes giving positive reaction out of			MPN index per 100 ml	95% confidence limits	
3 of 10 ml each	3 of 1 ml each	3 of 0.1 ml each		Lower	Upper
0	0	1	3	0.5	9
0	1	0	3	0.5	13
1	0	0	4	0.5	20
1	0	1	7	1	21
1	1	0	7	1	23
1	1	1	11	3	36
1	2	0	11	3	36
2	0	0	9	1	36
2	0	1	14	3	37
2	1	0	15	3	44
2	1	1	20	7	89
2	2	0	21	4	47
2	2	1	28	10	150
3	0	0	23	4	120
3	0	1	39	7	130
3	0	2	64	15	380
3	1	0	43	7	210
3	1	1	75	14	230
3	1	2	120	30	380
3	2	0	93	15	380
3	2	1	150	30	440
3	2	2	210	35	470
3	3	0	240	36	1300
3	3	1	460	71	2400
3	3	2	1100	150	4800

*The MPN tables reproduced here are included with the express permission of the American Public Health Association. The Committee expresses its gratitude for this privilege.

appropriate to the number of positive tubes for each dilution. For example, if in the illustration given in step 6 above, all positive tubes in the three selected dilutions (1:100, 1:1000, and 1:10,000) yielded confirmatory results for coliform organisms, then the values for each dilution are 3, 1, and 0 respectively. The table shows that this result indicates an MPN of 43 per 100 ml. To obtain the MPN of coliform organisms per gram of food, use the following formula:

$$\frac{\text{MPN from table}}{100} \times \text{dilution factor of middle tube} = \text{MPN/g.}$$

In the illustration given, this becomes

$$\frac{43}{100} \times 1000 = \text{MPN of 430 coliform organisms per g.}$$

Method 2[11]

I. Apparatus and materials

1   Requirements for Preparation and Dilution of the Food Homogenate, Method 1 or 2 (p. 60).

2   Incubator, $37° \pm 1°$ C.

3   1-ml bacteriological pipettes.

4   MacConkey Broth (Medium 41, Part III), 10 ml volumes in 150 × 15 mm tubes containing inverted Durham fermentation tubes (75 × 10 mm).

II. Procedure

1   Prepare food sample by one of the procedures recommended in the section on Preparation and Dilution of the Food Homogenate (p. 60). All techniques of dilution should be the same. Materials remaining from dilution blanks employed in the determination of the plate count can be used.

2   Pipette 1 ml of each of the decimal dilutions of food homogenate into each of three separate tubes of MacConkey Broth.

3   Incubate tubes at $37 ° \pm 1°$ C for 24 and 48 hours.

4   After 24 hours, record tubes showing gas production.[12] Return tubes not displaying gas to incubator for an additional 24 hours.

5   After 48 hours record tubes showing gas production.[12] The formation

11/This method is used in many countries for the enumeration of coliforms in water (Ministry of Health and Ministry of Housing and Local Government, 1956).

12/If tests for faecal coliforms are to be made, select a number of gas-positive tubes and follow the procedure described in Method 2 of the section on Determination of Coliform Organisms of Faecal Origin below.

of gas after 24 or 48 hours of incubation is considered sufficient evidence of the presence of coliforms.

6 To determine the MPN proceed as follows. Determine the highest dilution in which all three tubes are positive for gas production and the next two higher dilutions. If this is not possible because none of the dilutions yielded three positive tubes or because further dilutions were not made beyond the one yielding three positive tubes, use the last three dilutions. Following the example described in Method 1 above, II, steps 6 and 9, and referring to Table 2, determine the MPN.

Method 3[13]

I. Apparatus and materials

1 The requirements for Method 2 above, I, items 1 through 3.

2 Brilliant-Green Lactose Bile Broth 2% (Medium 10, Part III), 10 ml volumes in 150×15 mm tubes containing inverted Durham fermentation tubes (75×10 mm).

3 Violet-Red Bile Agar (Medium 100, Part III) or Endo Agar (Medium 21, Part III), for plates.

II. Procedure

1 Prepare food samples by one of the two procedures recommended in the section on Preparation and Dilution of the Food Homogenate (p. 60). The techniques of dilution should be the same. Material remaining from dilution blanks employed in the determination of the plate count can be used.

2 Pipette 1 ml of each of the decimal dilutions of the food homogenate into each of three separate tubes of Brilliant-Green Lactose Bile Broth 2%.

3 Incubate tubes at 35–37° C for 24 and 48 hours.

4 After 24 hours, record tubes showing gas production.[14] Return tubes not displaying gas to incubator for an additional 24 hours.

5 After 48 hours, record tubes showing gas production.[14]

6 Select tubes for confirmation tests and MPN determination as described in Method 1 above, II, step 6.

7 Confirm that tubes of Brilliant-Green Bile Broth 2% selected in step

13/This method is used in numerous laboratories on the European continent.
14/If tests for faecal coliforms are to be made, select a number of gas-positive tubes and follow the procedure described in Method 2 of the section on Determination of Coliform Organisms of Faecal Origin below.

6 immediately above are positive for coliform organisms by streaking a loopful from each on the surfaces of a plate of either Violet-Red Bile Agar or Endo Agar. Incubate plates inverted at 35–37° C; examine Violet-Red Bile Agar plates after 24 hours and Endo Agar plates after 48 hours. The formation of dark red colonies with diameters greater than 0.5 mm on Violet-Red Bile Agar or the formation of red colonies surrounded by red haloes on Endo Agar confirms the presence of coliform organisms.

8 Record the number of tubes in each dilution that were confirmed as positive for coliform organisms.

9 Determine the MPN of coliforms as described in Method 1 above, II, steps 6 and 9.

DETERMINATION OF COLIFORM ORGANISMS OF FAECAL ORIGIN

As a rule, sufficient measurement of the sanitary quality of foods is accomplished by means of the coliform determination. In certain cases, however, it may be desirable to determine the probable origin of coliform bacteria isolated from food and to obtain an estimate of the proportion of the total coliform density representative of recent faecal contamination. The following procedures may be expected to differentiate between coliforms of faecal origin (intestines of warm-blooded animals) and coliforms from other sources, with reasonable accuracy.

Method 1[15]

I. Apparatus and materials

1 Inoculating needle, with a 3-mm-diameter loop, preferably of nichrome or platinum-iridium wire.

2 Agitated water bath with thermo-regulation capable of maintaining a temperature of 45.5° ± 0.2° C.

3 E.C. Broth (Medium 19, Part III), 10 ml volumes in 150 × 15 ml tubes containing inverted Durham fermentation tubes (75 × 10 mm).

II. Procedure

1 Select tubes of Lauryl Sulphate Tryptose Broth that are positive for gas production in Method 1 of the section on Enumeration of Coliforms above.

15/This method (modified Eijkman test; Eijkman, 1904) follows that recommended for water (APHA et al., 1965) and foods (AFDOUS, 1966; Lewis and Angelotti, 1964).

2   Inoculate a loopful of broth from each of the selected cultures into a separate tube of E.C. Broth.

3   Incubate E.C. Broth tubes at $45.5° \pm 0.2°$ C and read for gas production after 24 and 48 hours.

4   E.C. Broth tubes displaying gas production may be presumed to be positive for faecal coliform organisms.

Method 2[16]

I. Apparatus and materials

1   Inoculating needle with a 3-mm-diameter loop, preferably of nichrome or platinum-iridium wire.

2   Agitated water bath with thermo-regulation capable of maintaining a temperature of $44° \pm 0.1°$ C.

3   5-ml bacteriological pipettes.

4   Brilliant-Green Lactose Bile Broth 2% (Medium 10, Part III), 10 ml volumes in 150 × 15 mm tubes containing inverted Durham fermentation tubes (75 × 10 mm).

5   Peptone Water (Medium 59, Part III), 10 ml volumes in 150 × 15 mm tubes.

6   Indole Reagent (Reagent 3, Part III).

II. Procedure

1   Select tubes of MacConkey Broth that are positive for gas production in Method 2 of the section on Enumeration of Coliforms above.

2   Inoculate a loopful of broth from each selected culture into a tube of Brilliant-Green Lactose Bile Broth 2% and a tube of Peptone Water.

3   Incubate tubes at $44° \pm 0.1°$ C.

4   Read the Brilliant-Green Lactose Bile Broth 2% tubes for gas production after 24 and 48 hours of incubation.

5   After 24 hours of incubation, pipette aseptically a 5 ml portion of each Peptone Water tube to a separate test tube and add 0.2–0.3 ml of Indole Reagent. Shake the tubes and let them stand for 10 minutes. A dark red colour in the amyl alcohol surface layer constitutes a positive indole test; the original colour of the reagent represents a negative test. Cultures showing gas production in Brilliant-Green Lactose Bile Broth 2% and

16/This method is used widely in several European countries in conjunction with Method 2 of the section on Enumeration of Coliforms above. As described, it follows the procedure outlined by Mackenzie *et al.* (1948).

indole fermentation in Peptone Water may be presumed to be positive for faecal coliform organisms.

IDENTIFICATION TESTS FOR COLIFORM ORGANISMS: THE IMViC PATTERN[17]

In routine analyses of foods, further confirmation of the presence of *E. coli* beyond that obtained from the tests described in Methods 1 and 2 of the section on Determination of Coliform Organisms of Faecal Origin above is not usually feasible because of the time and labour involved. In special instances, however, where the extra effort is merited, the differentiation of the coliform group into species and varieties can be carried out on the basis of the results of four tests (indole, methyl red, Voges-Proskauer, and sodium citrate) referred to collectively as the "IMViC tests." A grouping of reaction combinations is presented in Table 3.

TABLE 3*

Differentiation of coliforms

(All of these organisms are capable of producing acid and gas from lactose in 48 hours at 37 °C)

	Gas in lactose bile-salt medium at 44–45.5° C	Indole test	Methyl red test	Voges-Proskauer test	Growth in citrate
Escherichia coli					
Type I (typical)	+	+	+	−	−
Type II	−	−	+	−	−
Intermediates					
Type I	−	−	+	−†	+
Type II	−	+	+	−†	+
Enterobacter‡ aerogenes					
Type I	−	−	−	+	+
Type II	−	+	−	+	+
Enterobacter cloacae	−	−	−	+	+
Irregular					
Type I	−	+	+	−	−
Type II	+	−	+	−	−
Type VI	+	−	−	+	+
Irregular, other types		Reactions variable			

*This table is similar to the one given in Ministry of Health ... (1956).
†Weak positive reactions are occasionally found.
‡The Judicial Commission of the International Committee on Nomenclature of Bacteria of IAMS has officially substituted the term *Enterobacter* as the generic designation for the genus previously known as *Aerobacter* (Enterbacteriaceae Subcommittee, 1963).

17/The procedure for the IMViC tests follows closely those described in *Standard Methods for the Examination of Water and Wastewater* (APHA *et al.*, 1965).

I. Apparatus and materials

The media listed below include acceptable alternatives for the IMViC tests and for the purification of cultures. Decide on the procedures to follow before preparing the media.

1   Incubator, 35–37° C.

2   1-ml bacteriological pipettes with subdivisions of 0.1 ml or less.

3   Inoculating needle, preferably with nichrome or platinum-iridium wire.

4   Tryptone Broth (Medium 93, Part III), 5 ml. volumes in tubes.

5   Peptone Water (Medium 59, Part III), 5 ml volumes in tubes.

6   Buffered Glucose Broth (Medium 14, Part III), 10 ml volumes (methyl red test) and 5 ml volumes (Voges-Proskauer test) in tubes.

7   Salt Peptone Glucose Broth (Medium 71, Part III), 5 ml volumes in tubes.

8   Koser Citrate Broth (Medium 33, Part III), 5 ml volumes in tubes.

9   Simmons Citrate Agar (Medium 74, Part III), slants with 1 inch butts in tubes.

10   Eosin Methylene-Blue Agar (Medium 23, Part III), for plates.

11   Endo Agar (Medium 21, Part III), for plates.

12   Nutrient Agar (Medium 46, Part III), slants in tubes and supply for plates.

13   Lactose Broth (Medium 34, Part III), 10 ml volumes in 150 × 15 mm tubes containing inverted Durham fermentation tubes (75 × 10 mm).

14   Indole Reagent (Reagent 3, Part III).

15   Methyl Red Solution (Reagent 6, Part III).

16   Naphthol Solution (Reagent 7, Part III).

17   Potassium hydroxide solution (40% aqueous).

18   Materials for staining smears by Gram's method (Reagent 1, Part III).

II. Procedure

A. ISOLATION AND PURIFICATION OF CULTURES

1   Streak a loopful of each gas-positive broth tube from either method 1 (E.C. Broth) or Method 2 (Brilliant-Green Lactose Bile Broth 2%) of the section on Determination of Coliform Organisms of Faecal Origin (above) on a separate plate of Eosin Methylene-Blue Agar or of Endo Agar. Incubate plates inverted for 24 hours at 35–37° C.

2 Fish a representative colony (nucleated, with or without metallic sheen) from each plate and streak cells onto a Nutrient Agar plate. Incubate the plate inverted for 24 hours at 35–37° C.

3 Select individual colonies and transfer cells of each to a separate slant of Nutrient Agar and to a tube of Lactose Broth. Incubate cultures 24 hours at 35–37° C.

4 From those cultures producing gas in Lactose Broth, examine a smear stained by Gram's method to confirm the presence of Gram-negative non-spore-forming rods.

5 Use inoculating needle and 24-hour growth on slants of Nutrient Agar to inoculate IMViC media below.

B. INDOLE TEST (Kovacs, 1928)

1 Inoculate tubes of Tryptone Broth or Peptone Water from pure cultures. Incubate tubes at 35–37° C for 24 hours.

2 Add 0.2–0.3 ml of Indole Reagent to each tube and shake.

3 Let tubes stand for 10 minutes and observe results. A dark red colour in the amyl alcohol surface layer constitutes a positive test. An orange colour probably indicates the presence of skatole and may be reported as a ± reaction.

C. METHYL RED TEST (Ljutov, 1961)

1 Inoculate tubes of Buffered Glucose Broth from pure cultures. Incubate tubes at 35–37° C for 5 days.

2 Pipette 5 ml from each culture to a separate empty culture tube and add 5 drops of methyl red solution and shake.

3 Record a distinct red colour as methyl red positive, a distinct yellow colour as methyl red negative, and a mixed shade as questionable.

D. VOGES-PROSKAUER TEST (Ljutov, 1963; Levine, 1916)

1 Inoculate tubes of either Buffered Glucose Broth or Salt Peptone Glucose Broth from pure cultures and incubate tubes at 35–37° C. for 48 hours.

2 Pipette 1 ml of each culture to a separate empty culture tube and add 0.6 ml of naphthol solution and 0.2 ml of potassium hydroxide solution.

3 Shake the tubes, let them stand 2–4 hours, and observe results. Record the development of a pink to crimson colour in the mixture as a positive test.

E SODIUM CITRATE TEST (Koser, 1923; Simmons, 1926)

1 Inoculate tubes of Koser Citrate Broth or of Simmons Citrate Agar with cells from pure cultures. Use a straight needle and a light inoculum, since transfer of nutrients with the inoculum can invalidate the test. If Simmons Citrate Agar slants are used, stab the butts and streak the surfaces.

2   Incubate Koser Citrate Broth at 35–37° C for 72–96 hours. Incubate Simmons Citrate Agar at 35–37° C for 48 hours.

3   For both media, record visible growth as a positive reaction and no visible growth as a negative reaction. Growth is usually indicated by a change in the colour of the medium from light green to blue.

Enterobacteriaceae[18]

Collective enumeration of all members of the family Enterobacteriaceae has been recommended as a means of indicating enteric contamination, particularly for foods which have received a treatment that will have destroyed *E. coli* but not necessarily all salmonellae or other Enterobacteriaceae. This group of bacteria has not received widespread attention, and only a single method is being recommended for study. The Committee has included this method with a view to encouraging further experimentation to establish the significance of the organisms recovered and to determine if their use as indicators is warranted.

DETERMINATION OF MOST PROBABLE NUMBER

I. Apparatus and materials

1   Petri dishes, glass, 150×15 mm size preferable.

2   Pipettes: 1-ml bacteriological pipettes with subdivisions of 0.1 ml or less; 10-ml bacteriological pipettes.

3   Water bath or air incubator for tempering agar, $45° \pm 1°$ C.

4   Incubator, 35–37° C.

5   Inoculating needle, preferably with nichrome or platinum-iridium wire.

6   Enterobacteriaceae Enrichment Broth (Medium 22, Part III), 12 ml volumes in tubes and 100 ml volumes in 250 ml conical flasks.

7   Violet-Red Bile Glucose Agar (Medium 101, Part III), for plates.

II. Procedure

1   Prepare food sample by one of the two methods described in the section on Preparation and Dilution of the Food Homogenate (p. 60). Prepare only a 10^{-1} dilution of original homogenate.

18/The medium and method recommended follow those described by Mossel *et al.* (1962, 1963).

2 Pipette 10 ml of undiluted food homogenate to each of five flasks containing 100 ml each of Enterobacteriaceae Enrichment Broth.

3 Pipette 1 ml of undiluted food homogenate to each of five tubes containing 12 ml each of Enterobacteriaceae Broth.

4 Repeat step 3 using 1 ml of 10^{-1} dilution of food homogenate.

5 Mix tubes well and incubate at 35–37° C for 20–24 hours.

6 Shake the cultures and transfer one loopful of each to the surface of a separate plate of Violet-Red Bile Glucose Agar. Streak radially from the centre to the edge of the plate; 10–15 streaks, each 5–7 cm in length, should be made for each loopful.

7 Incubate plates inverted at 35–37° C for 20–24 hours and read and record results. Enrichment cultures are considered positive for Enterobacteriaceae if the streaked colonies are surrounded by a zone of precipitation accompanied by a purple discolouration of the agar.

8 Calculate the presumptive number of viable Enterobacteriaceae per gram of sample using the MPN table (Table 4).

CONFIRMATION TESTS FOR ENTEROBACTERIACEAE

As a rule, sufficient evidence of the presence of Enterobacteriaceae is obtained when positive results occur in the MPN tests above. In certain cases, however, it may be desirable to confirm that the colonies on the Violet-Red Bile Glucose Agar plates are Enterobacteriaceae. If so, follow the procedures outlined below.

I. Apparatus and materials

1 Requirements for Determination of Most Probable Number above, I, items 1 through 5 and 7.

2 Filter paper, Whatman No. 2, pieces 6 cm square.

3 Petrolatum, sterilized.

4 Nutrient Agar (Medium 46, Part III), slants in tubes.

5 Glucose Salt Medium (Medium 28, Part III), 3 cm depth in tubes.

6 Tetramethylparaphenylenediamine dihydrochloride, 1% aqueous solution.

7 Sulphanilic Acid Solution (Reagent 9, Part III).

8 α-Naphthylamine Solution (Reagent 8, Part III).

9 Materials for staining smears by Gram's method (Reagent 1, Part III).

TABLE 4

MPN index and 95% confidence limits for various combinations of positive and negative results when five 10-ml portions, five 1-ml portions, and five 0.1-ml portions are used

Number of tubes giving positive reaction out of			MPN index per 100 ml	95% confidence limits	
5 of 10 ml each	5 of 1 ml each	5 of 0.1 ml each		Lower	Upper
0	0	1	2	<0.5	7
0	1	0	2	<0.5	7
0	2	0	4	<0.5	11
1	0	0	2	<0.5	7
1	0	1	4	<0.5	11
1	1	0	4	<0.5	11
1	1	1	6	<0.5	15
1	2	0	6	<0.5	15
2	0	0	5	<0.5	13
2	0	1	7	1	17
2	1	0	7	1	17
2	1	1	9	2	21
2	2	0	9	2	21
2	3	0	12	3	28
3	0	0	8	1	19
3	0	1	11	2	25
3	1	0	11	2	25
3	1	1	14	4	34
3	2	0	14	4	34
3	2	1	17	5	46
3	3	0	17	5	46
4	0	0	13	3	31
4	0	1	17	5	46
4	1	0	17	5	46
4	1	1	21	7	63
4	1	2	26	9	78
4	2	0	22	7	67
4	2	1	26	9	78
4	3	0	27	9	80
4	3	1	33	11	93
4	4	0	34	12	93
5	0	0	23	7	70
5	0	1	31	11	89
5	0	2	43	15	114
5	1	0	33	11	93
5	1	1	46	16	120
5	1	2	63	21	150
5	2	0	49	17	130
5	2	1	70	23	170
5	2	2	94	28	220
5	3	0	79	25	190
5	3	1	109	31	250
5	3	2	141	37	340
5	3	3	175	44	500
5	4	0	130	35	300
5	4	1	172	43	490
5	4	2	221	57	700
5	4	3	278	90	850
5	4	4	345	120	1000
5	5	0	240	68	750
5	5	1	348	120	1000
5	5	2	542	180	1400
5	5	3	918	300	3200
5	5	4	1609	640	5800

II. Procedure

A. PURIFICATION OF CULTURES

1 Choose from the Violet-Red Bile Glucose Agar plates (Determination of MPN above, II, step 7) strictly at random, a number of discrete and apparently positive colonies.

2 Fish cells from each colony with an inoculating needle and streak a fresh separate plate of Violet-Red Bile Glucose Agar. Incubate plates inverted at 35–37° C for 24 hours.

3 Transfer cells from typical well-isolated colonies to slopes of Nutrient Agar and incubate tubes at 35–37° C for 24 hours.

4 Prepare smears of each culture and stain by Gram's method to confirm uniform cultures of Gram-negative rods.

5 Use cells from 24-hour slope cultures to inoculate media in tests described below.

B. OXIDASE TEST (Kovacs, 1956; Soc. Appl. Bact., 1966)

1 Place a 6-cm-square piece of filter paper in an empty Petri dish and add 3 drops of tetramethylparaphenylenediamine dihydrochloride solution to the centre.

2 With a platinum needle smear cells thoroughly on the reagent-impregnated paper in a line 3–6 mm long. Observe the colour. The oxidase test is positive if transferred cells turn dark purple in 5–10 seconds. Enterobacteriaceae are oxidase-negative.

C. GLUCOSE FERMENTATION

1 Stab two tubes of Glucose Salt Agar for each culture. Use a light inoculum.

2 Cover the agar in one tube with a layer of melted petrolatum to a depth of ¼–1 cm.

3 Incubate tubes at 35–37° C for 48 hours and record results. The test is positive when an acid reaction (yellow colour) occurs in both tubes. It is negative when the acid reaction occurs only in the open tube. Enterobacteriaceae give a positive reaction.

Faecal streptococci (Lancefield group D streptococci)

The enterococci comprise the species *Streptococcus faecalis* (and its varieties) and *Strep. faecium*. Together with *Strep. bovis* and *Strep. equinus*, they represent the group of streptococci serologically defined as Lancefield group D. The whole group is loosely called faecal streptococci, because different members are normally present in mammalian faeces, and for this reason they have been used as indicators of faecal contamination.

The faecal streptococci in foods are usually enumerated as an aid to the estimation of the sanitation practices in food factories, particularly regarding inadequate cleansing and disinfection of factory surfaces; to support other indications of faecal contamination; and as an index of general microbiological quality. Their capacity to cause food-poisoning is still in doubt.

There is also continuing doubt about the significance of the pathological and ecological differences between the several species. Hence, the Committee has recommended first a method which would detect all of them, rather than the more-limited, more-resitant group representing the enterococci according to Sherman's original definition (Sherman, 1937).

The procedure considered to be the more generally satisfactory involves three stages: (1) presumptive enumeration using Packer's Crystal-Violet Azide Blood Agar (Packer, 1943) and the pour plate technique; (2) confirmatory identifications of suspect colonies after purification on Thallous Acetate Tetrazolium Glucose Agar (Barnes, 1956); (3) identification of species.

This procedure begins with partial selection in a not highly selective medium, in which the selective agents are crystal violet and azide. Among the colonies which grow on this medium, those of the group D streptococci are usually distinguished by their small size and purple colour. The confirmatory tests make use of the ability of the group D streptococci to grow at 45° C, of the ability of these faecal organisms to resist bile, and of their lack of catalase.

Identification of species requires the standard taxonomic tests, but may sometimes be necessary after following the above procedure, because *Strep. bovis* and *Strep. equinus*, unlike the enterococci *Strep. faecalis* and *Strep. faecium*, do not possess unusual resistance to heating, freezing, drying, and disinfectants, and because different species may preponderate in different animals (cf. Part I, p. 29).

When one wishes to concentrate on the more-resistant enterococci, the medium of Barnes (1956) may be used in the enumeration procedure. This medium depends on the strongly selective action of thallous acetate, and tends to favour *Strep. faecalis* (and its varieties), which produces red colonies, and *Strep. faecium*, which produces white. This method requires less work for the identification of colonies.

ENUMERATION OF PRESUMPTIVE FAECAL STREPTOCOCCI

I. Apparatus and materials

1 Petri dishes, glass (100 × 15 mm) or plastic (90 × 15 mm).

2 1-ml bacteriological pipettes.

3 Water bath or air incubator for tempering agar, 45° ± 1° C.

4 Incubator, 35–37° C.

5 Packer's Crystal-Violet Azide Blood Agar (Medium 51, Part III), for plates.

6 Colony counter (Quebec dark field or equivalent type recommended).

7 Tally register.

II. Procedure

1 Prepare food samples by one of the two methods described in the section on Preparation and Dilution of the Food Homogenate (p. 60), using the same techniques of dilution. Materials remaining from dilution blanks employed in the determination of the plate count can be used.

2 To duplicate sets of Petri dishes, pipette 1 ml aliquots of each of the decimal dilutions of the food homogenate.

3 Promptly add to each dish 15 ml of Packer's Crystal-Violet Azide Blood Agar, melted and tempered to 45° ± 1° C.

4 Mix the samples and agar by rotating and tilting the dishes.

5 After the agar has solidified, incubate the plates inverted at 35–37° C for 72 hours.

6 Using the colony counter and tally register, count all the small violet-coloured colonies on plates containing 30–300 colonies.

7 Compute the number of presumptive faecal streptococci per gram of food specimen.

CONFIRMATION OF FAECAL STREPTOCOCCI

I. Apparatus and materials

1 The requirements listed for Enumeration of Presumptive Faecal Streptococci immediately above, I, items 1 through 4.

2 Oven or incubator for drying agar plates.

3 Inoculating needle, preferably with nichrome or platinum-iridium wire.

4 Thallous Acetate Tetrazolium Glucose Agar (Medium 86, Part III), for plates.

5 Tryptose Agar (Medium 91, Part III), slants in tubes.

6 Tryptose Bile Broth 40% (Medium 92, Part III), 5 ml volumes in tubes.

7 Hydrogen peroxide, 25–30% aqueous solution.

8 Materials for staining smears by Gram's method (Reagent 1, Part III).

II. Procedure

A. ISOLATION AND PURIFICATION OF CULTURES

1 Prepare plates of Thallous Acetate Tetrazolium Glucose Agar and allow plate surfaces to dry (see item 9, p. 158, for note on drying plates).

2 Select three to five colonies from each type of violet-coloured colony on the plates from section on Enumeration of Presumptive Faecal Streptococci above.

3 With an inoculating needle streak cells from each selected colony on a separate dried plate of Thallous Acetate Tetrazolium Glucose Agar. Incubate plates inverted at 35–37° C for 48 hours.

4 Select two white and two red-centred colonies; do not select intensely red colonies (*Strep. lactis* type). Colonies with red centres are probably *Strep. faecalis*, whereas white colonies are likely to be *Strep. faecium* or other Lancefield group D streptococci with low reducing powers.

5 Transfer cells from selected colonies to separate slants of Tryptose Agar. Incubate at 35–37° C for 24 hours.

6 Prepare a smear of each culture and stain by Gram's method to confirm the presence of Gram-positive cocci.

7 Use cells of 24-hour slant cultures to inoculate for tests described in B, C, and D below.

B. GROWTH ON TRYPTOSE AGAR

1 Inoculate slopes of Tryptose Agar from pure cultures and incubate at 45° ± 0.1° C in a water bath for 48 hours.

2 Record results; growth indicates a positive test.

C. GROWTH IN TRYPTOSE BILE BROTH 40%

1 Inoculate tubes of Tryptose Bile Broth 40% from pure cultures and incubate at 35–37° C for 48 hours.

2 Record results; growth (turbidity) represents a positive test.

D. CATALASE REACTION

1 Inoculate slants of Tryptose Agar from pure cultures and incubate at 35–37° C for 48 hours.

2 Pipette sufficient hydrogen peroxide solution to the tubes to cover the growth on the slants.

3 Hold tubes at eye level and observe. The evolution of bubbles indicates a positive test.

Cultures which grow on Tryptose Agar at 45° C and in Tryptose Bile Broth 40% and are negative for catalase are considered confirmed faecal streptococci.

IDENTIFICATION OF SPECIES

I. Apparatus and materials

1 The requirements listed in the section on Confirmation of Faecal Streptococci immediately above, I, items 1 through 3.

2 Tryptose Broth, pH 9.6 (Medium 94, Part III), 5 ml volumes in tubes.

3 Tryptose Salt Broth (Medium 95, Part III), 5 ml volumes in tubes.

4 Phenol-Red Sorbitol Broth (Medium 60, Part III), 5 ml volumes in tubes.

5 Tryptose Tellurite Agar (Medium 98, Part III), slants in tubes.

6 Tryptose TTC Agar (Medium 99, Part III), slants in tubes.

7 Tryptose Broth (Medium 94, Part III), adjusted to pH 7.2.

8 Water bath with thermo-regulator, 60° ±1° C.

II. Procedure

1 Using inoculating needle inoculate two tubes of the following media with 24-hour-old cells of each culture examined; incubate at 35–37° C for 48 hours; growth represents a positive reaction: (i) Tryptose Broth, pH 9.6. (ii) Tryptose Salt Broth. (iii) Tryptose Tellurite Agar slants. (iv) Tryptose TTC Agar slants.

2   Inoculate two tubes of Phenol-Red Sorbitol Broth and incubate at 35–37° C for 48 hours. The development of a yellow colour in the medium indicates fermentation and a positive reaction.

3   For each culture, preheat two tubes of Tryptose Broth (pH 7.2) to 60° C. Inoculate both tubes liberally and hold them at 60° C for 30 minutes in a water bath. Cool tubes under tap and incubate them for 48 hours at 35–37° C. Growth is a positive test.

4   Classify cultures according to scheme in Table 5.

TABLE 5

Identification of Lancefield group D streptococci
(Growth on different media and under different test conditions)

Media or test conditions	*Strep. faecalis*	*Strep. faecium*	*Strep. durans*	*Strep. bovis*	*Strep. equinus*
45° C	+	+	+	+	+
40% bile	+	+	+	+	+
Catalase	−	−	−	−	−
pH 9.6	+	+	+	−	−
6.5% NaCl	+	+	+	−	−
Survival at 60° C for 30 minutes	+	+	+	−	−
0.05% tellurite	+	−	−	−	−
Sorbitol	+	±	−	−	−
TTC Agar	+	−	±	±	−

Salmonellae

All salmonellae should be considered as potential pathogens of man. As discussed in Part I, p. 9, no single method has yet been developed with adequate sensitivity for the recovery of diverse serotypes of salmonellae from all types of foods. The procedures recommended here by the Committee, therefore, take into account specific requirements which experience has shown are effective for different categories of foods.

Methods for the isolation and identification of salmonellae from foods may consist of a sequence of six stages:

  1 Non-selective enrichment culture.

  2 Selective enrichment culture.

  3 The use of selective agar media.

4 Verification of selected colonies by use of determinative biochemical tests.

5 Serological recognition in two steps: (a) the use of preparations of polyvalent O and polyvalent H antisera, and (b) typing by use of O grouping antisera and H pooled antisera.

6 Typing by means of bacteriophage where available.

Stages 1 to 3 may be referred to as the isolation procedure, and stages 4 to 6 as the identification procedure. The identification procedure is essentially the same no matter what the source of the culture being examined, subject only to minor local preferences, which are not critical. Stages 1, 2, and 3 are subject to change depending on the nature of the test food, and correct use is critical for effective recovery, particularly when salmonellae are few and competitive organisms are many.

Non-selective enrichment is normally used only when the test food either has undergone a process such as heating, drying, and irradiation or has been exposed to extended freezing, or has a low pH. These treatments or conditions may impart a semidormant or "weakened" state to any *Salmonella* cells present. The purpose of non-selective enrichment is to allow the cells to begin normal growth processes without being exposed to inhibitory or selective substances which might be unduly toxic to the cell in a "weakened" condition. Selective enrichment media contain such inhibitory substances.

The use of an enrichment broth is an attempt to encourage the multiplication of salmonellae while reducing or inhibiting growth of competitive organisms, such as coliforms, *Proteus* species, and pseudomonads, which otherwise might outgrow any salmonellae, particularly if the ratio of competitors to salmonellae is high. However, since there is some evidence to suggest that different salmonellae demonstrate different sensitivities to the inhibitory substances in selective broths, and since the nature and extent of the background flora in a given food are not always known, the most suitable selective media for each particular circumstance cannot at present be specified with certainty. In practice there is a great deal of variation in the preferences of individual laboratories. Nonetheless, for some food categories the background flora and the modifying effect of the food material on the selective action of the inhibitors are known from experience, and, therefore, in these cases, specific recommendations can be made. Generally, however, the use of two different selective enrichment broths is advised for each test.

Selective agars usually contain inhibitors and also an indicator system which specifically colours particular types of colonies or imparts a colour to the agar surrounding the colony, thus allowing recognition of suspect

Salmonella colonies. Six different selective agar media in frequent use in laboratories around the world are recommended, in some cases specifically and in others as alternatives depending largely on the type and number of competing organisms likely to be associated with the food material in question. These media can be divided into two categories based on their relative inhibitory action on Gram-negative organisms:

Category 1   Brilliant Green Agar
                    Brilliant-Green Sulphadiazine Agar
                    Brilliant-Green MacConkey Agar

Category 2   Desoxycholate Citrate Agar
                    Salmonella-Shigella Agar
                    Bismuth Sulphite Agar

The six media represent a range of selective power with those in category 2 being generally more inhibitory for Gram-negative organisms. It is recommended, without reference to specific foods, that where the nature of the competing flora is unknown or varied, one medium from each category be used.

An adequate sample size is important for the examination of all foods for salmonellae. The procedures described below are based on a minimum sample of 25 g.

METHODS FOR ISOLATION OF SALMONELLAE

This part includes three main stages: (i) non-selective enrichment, (ii) selective enrichment, and (iii) plating on selective agar media. As described above, the non-selective enrichment step is omitted for some foods. The procedures for these stages are given separately for three different food categories.

I. Apparatus and materials

Note that the materials required for all food categories are included under this heading. Therefore, not all of the enrichment broths and selective agar media listed are required for any one food. Follow selection recommendations given in the procedure or, if preferred, select other combinations of these media.

1   Mechanical blendor, operating at not less than 8000 rpm and not more than 45,000 rpm.

2   Glass or metal blending jars of 1 litre capacity, with covers, resistant to autoclave temperatures; one jar for each food specimen to be analysed.

3 Balance with weights, capacity at least 2500 g, sensitivity 0.1 g. (A large laboratory torsion balance meets these specifications.)

4 Instruments for preparing samples: knives, forks, forceps, scissors, spoons, spatulas, sterilized prior to use by autoclaving or in an oven.

5 Incubator, 35–37° C.

6 Incubator or drying oven for drying plates (see item 9, p. 158, for note on drying plates).

7 Water baths with thermo-regulators: one at 45° ± 1° C, for tempering agar; one 43° ± 0.2° C, for incubating enrichment cultures.

8 Conical flasks or screw-capped glass jar, approximate capacity 500 ml.

9 1-ml bacteriological pipettes.

10 Inoculating needle with nichrome or platinum-iridium wire.

11 Petri dishes, glass (100 × 15 mm) or plastic (90 × 15 mm).

12 Lactose Broth (Medium 34, Part III), bulk.

13 Selenite Cystine Broth (Medium 72, Part III), bulk and in 10 ml volumes in tubes as required.

14 Tetrathionate Broth (Medium 81, Part III).

15 Tetrathionate Brilliant-Green Bile Broth, Kauffmann modification (Medium 83, Part III), bulk and in 10 ml volumes in tubes as required.

16 Tetrathionate Brilliant-Green Broth, Rolfe modification (Medium 82, Part III), bulk and in 10 ml volumes in tubes as required.

17 Tetrathionate Brilliant-Green Bile Sulphathiazole Broth (Medium 84, Part III), bulk and in 10 ml volumes in tubes as required.

18 Tetrathionate Novobiocin Broth (Medium 85, Part III), bulk and in 10 ml volumes in tubes as required.

19 Brilliant Green Agar (Medium 9, Part III), for plates.

20 Brilliant-Green Sulphadiazine Agar (Medium 12, Part III), for plates.

21 Brilliant-Green MacConkey Agar (Medium 11, Part III), for plates.

22 Desoxycholate Citrate Agar (Medium 18, Part III), for plates.

23 Salmonella-Shigella Agar (Medium 66, Part III), for plates.

24 Bismuth Sulphite Agar (Medium 2 or 3, Part III), for plates.

II. Procedure for dried egg products, frozen pasteurized liquid eggs, and frozen dried processed foods

A. NON-SELECTIVE ENRICHMENT

1 Thoroughly mix each of two 25 g samples of test material with 225 ml of Lactose Broth in separate screw-capped jars or conical flasks (approx. capacity 500 ml).[19]

2 Incubate for 24–48 hours at 35–37° C, and proceed with selective enrichment.

3 Also streak a loopful of each incubated culture onto selective agar as described in c below.

B. SELECTIVE ENRICHMENT

1 Pipette 1 ml of non-selective enrichment culture to a 10 ml tube of Selenite Cystine Broth and 1 ml to a 10 ml tube of one of the five tetra-thionate broths, Media 81–85, Part III (the choice will depend on the individual preference of the laboratory).

2 Incubate broth tubes 24 hours at 35–37° C or at 43° C. If both temperatures are used, prepare duplicate subcultures for each selective enrichment medium and proceed with plating on selective agar media, c below.

C. PLATING ON SELECTIVE AGAR MEDIA

1 Prepare dried plates of two selective agar media, one medium from each of the two categories of selective media listed in the introduction to this section on Salmonellae. Generally, with these foods, either Brilliant Green Agar or Brilliant-Green Sulphadiazine Agar is used.

2 Transfer a 5 mm loopful of each enrichment broth culture to the surface of one plate each of the two selective agar media, and spread in a manner to obtain isolated colonies.

3 Incubate plates inverted at 35–37° C for 24 hours. If typical colonies do not appear after this time, reincubate and examine again after 48 hours.

(*a*) Typical *Salmonella* colonies on Brilliant Green Agar and on Brilliant-Green Sulphadiazine Agar appear colourless, pink to fuchsia, or translucent to opaque with the surrounding medium pink to red. Some salmonellae appear as translucent green colonies when surrounded by lactose- or sucrose-fermenting organisms which produce colonies that are yellow-green or green in colour.

(*b*) Typical *Salmonella* colonies on Brilliant-Green MacConkey Agar appear colourless and transparent. Since competing coliform organisms precipitate the bile salts in the medium, *Salmonella* colonies growing in

19/If the concentration of sugar in the sample is more than 5%, dilute with additional broth so that the enriched sample contains less than 5% of sugar. Similarly, acid products should be adjusted to a neutral pH with more broth or with a dilute solution of sodium hydroxide.

proximity to coliform colonies will have the appearance of clearing the areas of precipitated bile.

(c) Typical *Salmonella* colonies on Salmonella-Shigella Agar appear uncoloured to pale pink, opaque, and translucent. Some strains produce black-centred colonies.

(d) Typical *Salmonella* colonies on Desoxycholate Citrate Agar appear flat with colourless edges and grey or black centres.

(e) Typical *Salmonella* colonies on Bismuth Sulphite Agar appear brown, grey, to black, sometimes with a metallic sheen. The medium surrounding the colony is usually brown at first, then turns black as the incubation time increases. Some strains produce green colonies with little or no darkening of the surrounding medium.

4 Select several suspect colonies from each selective agar medium used for the identification tests described in the section on Procedures for Identification of Salmonellae below. If the purpose of the examination is to determine the number and the relative proportion of the different sero-types present in the specimen, then as many as twelve (six from each selective agar) colonies could be selected. If the object is simply to determine the presence or absence of salmonellae, then only two typical colonies from each agar medium need be used in identification tests.

5 If the agar plates are crowded with coliform organisms, streak new plates of the chosen selective agar media using a 1:1000 dilution of the enrichment cultures. The enrichment cultures can be held at room temperature or in the refrigerator at 5–8° C during the time the original set of streaked plates are incubating.

6 Selective agar plates containing typical *Salmonella* colonies should be held at 5–8° C until identification tests with chosen colonies are completed.

III. Procedure for raw meat, poultry, and unpasteurized eggs

Non-selective enrichment is not required for these foods.

A. SELECTIVE ENRICHMENT

1 If the food is frozen, thaw a suitable portion of the specimen (it may be necessary to use aseptically sawn or drilled pieces) overnight at 5–10° C.

2 (a) Weigh 25 g of sample into each of two tared sterile jars (capacity approx. 500 ml), cut the sample into small pieces with scissors, and add 225 ml of one of the following broths to each jar: Selenite Cystine Broth or one of the five tetrathionate broths listed in I, items 14–18, above. The choice will usually depend on the experience with the particular food in question or on the preferences of individual laboratories.

(b) Incubate one jar for 24 hours at 43° ± 0.2° C and the duplicate jar for 24 hours at 35–37° C.

B. PLATING ON SELECTIVE AGAR MEDIA

Follow the procedure described under II, c above, steps 1 through 6. For this category of food all of the agar media in each of the two groups have been used. The choice will be based on the availability of media or ingredients or on the preferences of individual laboratories.

IV. Procedure for powdered milk

A. PRIMARY SELECTIVE ENRICHMENT

1   Reconstitute 100 g of powdered milk in 1 litre of sterile distilled water in a 2-litre flask. If the pH is below 6.6, adjust to 6.8 or slightly above.

2   Add 20 ml of 0.1% aqueous solution of brilliant green dye (final ratio of dye to milk, 1:50,000 (North, 1960)).

3   Incubate the flask at 35–37° C for 18–24 hours. Since this primary selective enrichment culture is streaked directly onto selective agar media as well as being subjected to a secondary enrichment procedure, proceed with B and C, immediately below, at the same time.

B. SECONDARY SELECTIVE ENRICHMENT

1   Pipette 10 ml of primary selective enrichment culture to 100 ml of one of the five tetrathionate broths listed in I above.

2   Incubate broth culture at 35–37° C for 18–24 hours and proceed with c immediately below.

C. PLATING ON SELECTIVE AGAR MEDIA

1   Prepare plates of Brilliant-Green Sulphadiazine Agar, following instructions given in item 9, p. 158, for drying surfaces.

2   Streak 1 loopful of each primary and secondary enrichment culture onto surface of one plate of Brilliant-Green Sulphadiazine Agar in a manner to obtain individual colonies.

3   Follow II, c, steps 1 through 6, above, regarding incubation and selection of colonies.

PROCEDURES FOR IDENTIFICATION OF SALMONELLAE

This section includes three main stages in the identification of suspected salmonellae cultures: (1) verification of selected colonies by use of determinative biochemical tests; (2) serological recognition by use of preparations of polyvalent somatic O, polyvalent H, O grouping, and H pooled antisera; and (3) typing by means of bacteriophage.

Laboratories concerned with the detection of salmonellae in foods normally use abbreviated biochemical and serological tests for the tentative determination of the serotype. Those laboratories having access to specialized national or state salmonella typing centres then send representative cultures to such centres for definitive typing.

Numerous biochemical tests can be used to assist in identification, but usually sufficient information is obtained with relatively few tests, involving mainly the use of one or two multiple reaction media, to warrant proceeding quickly with serological testing. Gillies Media 1 and 2 (determine motility; fermentation of glucose, mannitol, and sucrose or salicin; and production of urease, indole, and H_2S) or Triple Sugar Iron Agar (determines oxidation or fermentation of the sugars employed—sucrose, glucose, and lactose—and production of H_2S) are widely used in this respect.

Suspected *Salmonella* cultures, based on the biochemical reactions, are next tested with polyvalent O and polyvalent H antisera. These antisera contain antibodies collectively representing the majority of salmonellae. Cultures giving positive reactions with polyvalent antisera are further tested with O grouping and H pooled antisera. The latter preparations are composed of antibodies representative of the chief antigens common to particular groups of salmonellae, in accord with the Kauffmann-White scheme of identification (Edwards and Ewing, 1962). These "groups" are alphabetically identified A to I. Positive agglutination in this test indicates the group to which the culture belongs and the "probable" serotype. For definitive typing, specific O and H antisera are required.

Abbreviated methods recommended below may be summarized as follows: select typical colonies from selective agar plates; inoculate Gillies Media 1 and 2 or Triple Sugar Iron (TSI) Agar and Lysine Iron Agar; if cultures appear contaminated, purify on MacConkey Agar; test suspect salmonellae cultures with polyvalent O and polyvalent H antisera (reactions typical of salmonellae in Gillies Media 1 and 2 or in TSI and Lysine Iron Agar and positive agglutination indicate a probable *Salmonella*); then test cultures with O grouping sera and Spicer-Edwards pooled H antisera (positive agglutination indicates the group, A to I, and "probable" serotype); preferably, forward the culture to a reference laboratory for definitive typing.

Biochemical screening tests for salmonellae

Two test series are outlined below, Test Series A and Test Series B. Either can be used for the differential determination of presumptive salmonellae. They differ only in that Test Series A involves the use of Gillies Media 1 and 2 for this purpose and Test Series B involves the use of Triple Sugar Iron Agar and Lysine Iron Agar.

I. Apparatus and materials

1 Requirements listed above in the section on Methods for Isolation of Salmonellae, I, items 5 through 11.

2 Agitated water bath with thermo-regulation capable of maintaining a temperature of $37° \pm 0.5°$ C.

3 MacConkey Agar (Medium 40, Part III), slants and bulk for plates.

4 Nutrient Agar slants (Medium 46, Part III).

5 Gillies Medium 1 (Medium 25, Part III), slants with 25–30 mm butts in tubes.

6 Gillies Medium 2 (Medium 26, Part III), 50–60 mm depth in tubes, and Lead Acetate and Indole papers (Reagents 4 and 2, Part III).

7 Lysine Iron Agar (Medium 39, Part III), for plates.

8 Triple Sugar Iron Agar (TSI Agar; Medium 88, Part III; either formula 1 or formula 2), slants with 25–30 mm butts in tubes.

9 Materials for staining by Gram's method (Reagent 1, Part III).

II. Procedure for Test Series A

1 Purify colonies selected from agar plates referred to in the section on Methods for Isolation of Salmonellae above (see II, c, step 4 of that section for comment on number of colonies to choose). For practical purposes, if time is limited, culture purification may be skipped at this point and performed, if necessary, from Gillies Medium 1 or 2 below. Use the following procedure to purify cultures:

(*a*) Streak cells from each colony onto the surface of a separate plate of MacConkey Agar in such a manner as to obtain isolated colonies.

(*b*) Incubate plates inverted at 35–37° C for 24 hours.

(*c*) *Salmonella* colonies on MacConkey Agar plates appear transparent and colourless, sometimes with pinkish centres. If many lactose fermenters are present and bile has precipitated, a *Salmonella* colony will give the appearance of having cleared the precipitate in the area immediately surrounding the colony.

(d) Pick typical colonies to slants of Nutrient Agar and incubate cultures at 35–37° C for 24 hours.

(e) Prepare smears from agar slant cultures and stain by Gram's method. If cultures are pure, use to inoculate media below.

2 Determine presumptive salmonellae by inoculating each purified culture into one tube each of Gillies Media 1 and 2 as described below. Use as the inoculum cells from 24-hour-old Nutrient Agar slope cultures.

(a) *Gillies Medium 1*

(i) Inoculate with straight needle by streaking the slope and stabbing the butt. Incubate 24 hours at 35–37° C.

(ii) Record urease reaction, glucose and mannitol fermentation, and gas production. Urease-positive cultures turn medium dark purple. Mannitol fermentation is indicated by a yellow butt and gas production by formation of gas pockets in the agar. Typical salmonellae are urease-negative but ferment both glucose and mannitol with or without gas formation.

(iii) Discard all urease-positive cultures and all those that fail to ferment both glucose and mannitol.

(b) *Gillies Medium 2*

(i) Inoculate with straight needle by making a single stab approximately 2 cm deep into the medium with the tube held in an upright position.

(ii) Insert into each tube the Lead Acetate and Indole test papers as described under Reagents 4 and 2, Part III. Do not permit the papers to touch each other.

(iii) Incubate tubes in an upright position at 35–37° C for 24 hours. If all reactions (see (iv) below) are not clear, reincubate a further 24 hours.

(iv) Record sucrose or salicin fermentation, motility, and H_2S and indole production. A colour change in the agar from the original blue green to yellow indicates either sucrose or salicin fermentation or both. Motility is indicated by a diffuse zone of growth spreading from the line of inoculation. A darkening of the lead acetate paper shows hydrogen sulphide production. A reddish-coloured indole paper represents a positive test for indole formation. Typical salmonellae are motile, H_2S-positive, indole-negative, and negative for the fermentation of both sucrose and salicin.

(v) Discard indole-positive cultures.

3 If one or more cultures are eliminated by their reactions in Gillies Media 1 and 2, pick an equivalent number of suspect colonies from the corresponding selective agar plates and repeat steps 1 and 2 immediately

above. Note that this step is required only if sufficient presumptive positive cultures have not already been found. In cases where one of the
selective agar media fails to give any or a sufficient number of presumptive salmonellae, make up the desired complement of cultures with presumptive positive isolates from the other selective agar medium. This is
only important where the object is to determine the serotypes present
and their relative numbers.

4 Submit presumptive *Salmonella* cultures to serological tests described
under Serological Tests below.

III. Procedure for Test Series B

1 Purify suspect colonies from selective agar plates as described in II,
step 1, above. For practical purposes, if time is limited, this step may be
omitted at this point and performed, if necessary, from the differential
sugar medium after 2, (c), below.

2 Determine presumptive salmonellae as follows:

(*a*) Inoculate one tube each of Triple Sugar Iron Agar (TSI Agar) and
Lysine Iron Agar with 24-hour-old purified culture on Nutrient Agar or
directly from a single suspected colony on selective agar plates (see II, c,
step 4, of section on Methods for Isolation of Salmonellae above for comment on number of colonies to pick). Inoculate TSI Agar and Lysine
Iron Agar with needle by streaking the slant and stabbing the butt.

(*b*) Incubate the cultures overnight at 35–37° C.

(*c*) Discard cultures that do not give reactions typical of salmonellae in
the test media.[20] Typical reactions in TSI Agar are indicated by a red
slant (alkaline reaction) and a yellow butt (acid; glucose fermentation),
with or without production of H_2S and gas (H_2S indicated by blackening
of the medium). Typical reactions of salmonellae and *Arizona* species
on Lysine Iron Agar are indicated by a light purple slant and butt (alkaline reaction), with production of H_2S and sometimes gas. Cultures that
have not been purified should be streaked to MacConkey Agar Plates as
described above (II, step 1).

3 If all cultures are eliminated as a result of their action on TSI or
Lysine Iron Agar, pick additional colonies from selective agar plates and
screen.

4 Submit presumptive *Salmonella* cultures to serological tests described
under Serological Tests below.

20/Lactose-fermenting *Arizona* organisms and the rare lactose-fermenting salmonellae will be detected by use of Lysine Iron Agar; gas pockets will form in the agar.

Serological tests[21]

I. Apparatus and materials

1 Glass slides (2 × 3 in.) or Petri dishes (100 × 15 mm).

2 Inoculating needles and loops, preferably with platinum-iridium or nichrome wire.

3 H Broth (Medium 29, Part III), 4 ml volumes in 13 × 100 mm tubes.

4 Sodium chloride (0.85% aqueous solution).

5 Pure cultures of presumptive salmonellae on Triple Sugar Iron or Nutrient Agar slants and in H Broth as required.

6 Pure cultures of known *Salmonella* serotypes on Nutrient Agar slants and in H Broth.

7 Pipettes: 0.2 ml capacity with 0.01 ml gradations; 1.0 ml capacity with 0.01 ml gradations; and 5 ml and 10 ml sizes with 0.1 ml gradations.

8 Serological test tubes, 75 × 10 mm or 100 × 13 mm.

9 Thermo-regulated water bath, 50° ± 1° C.

10 0.6% formalized saline solution (0.85% saline solution containing 0.6 ml of formalin per 100 ml).

11 *Salmonella* antisera

(*a*) *Salmonella* polyvalent O (somatic) antiserum which contains agglutinins for at least the O antigens: 1–16, 19, 22, 23, 24, 25, and Vi.

(*b*) *Salmonella* individual O (somatic) antisera for at least each of the groups: A, B, C_1, C_2, D, E (E_1, E_2, E_3, E_4), F, G, H, I, and Vi.

(*c*) *Salmonella* polyvalent H (flagellar) antiserum containing agglutinins for at least the following H antigens: a, b, c, d, eh, en, enx, fg, fgt, gm, gmq, gms, gp, gpu, gq, gst, gt, i, k, lv, lw, lz_{13}, lz_{28}, mt, r, y, z, z_4z_{23}, z_4z_{24}, z_4z_{32}, z_6, z_{10}, z_{29}, 1, 2; 1, 5; 1, 6; and 1, 7.

(*d*) *Salmonella* Spicer-Edwards H (flagellar) antisera consisting of seven pooled antisera which react as follows:

(i) *Salmonella* H antiserum Spicer-Edwards *1* which reacts with antigens: a, b, c, d, eh, fg, fgt, gm, gms, gmt, gp, gpu, gq, gst, ms, mt, and i.

(ii) *Salmonella* H antiserum Spicer-Edwards *2* which reacts with antigens: a, b, c, k, r, y, and z_{29}.

(iii) *Salmonella* H antiserum Spicer-Edwards *3* which reacts with antigens: a, d, eh, k, z, z_4z_{23}, z_4z_{24}, z_4z_{32}, and z_{29}.

(iv) *Salmonella* H antiserum Spicer-Edwards *4* which reacts with antigens: b, d, fg, fgt, gm, gms, gmt, gp, gpu, gq, gst, ms, mt, k, r, z, and z_{10}.

21/Edwards and Ewing (1962); AOAC (1967).

(v) *Salmonella* H antiserum *e, n* complex which reacts with antigens: enx and enz$_{15}$.

(vi) *Salmonella* H antiserum *L* complex which reacts with antigens: lv, lw, lz$_{13}$, lz$_{28}$.

(vii) *Salmonella* H antiserum *1* complex which reacts with antigens: 1, 2; 1, 5; 1, 6; 1, 7; and z$_6$.

II. Procedure for polyvalent O (somatic) test (slide or plate test)

1   Dilute and adequately pretest antisera with known test cultures, to ensure reliability of test results with unknown cultures. Follow procedure given in steps 2 to 7 immediately below.

2   Using a wax pencil mark off two sections about 1 by 2 cm on the inside of a glass Petri dish or on a 2 × 3 inch glass slide.

3   Place a small amount (1.5 mm loopful) of culture from a Nutrient Agar or Triple Sugar Iron Agar slant (24 or 48 hours) directly on the dish or slide in the upper part of each marked section.

4   Add 1 drop of 0.85% sodium chloride solution to the lower part of each marked section. With a clean, sterile transfer loop or needle, emulsify the culture in the saline solution for one section and repeat for the other section.

5   Add a drop of *Salmonella* polyvalent O antiserum to one section of emulsified culture and mix with a sterile loop or needle.

6   Tilt the mixture in both sections back and forth for 1 minute and observe against a dark background. A positive reaction is indicated by a rapid, strong agglutination.

7   Classify the polyvalent O (somatic test) as:

(*a*) Positive when there is agglutination in the culture-saline-serum mixture and no agglutination in the culture-saline mixture.

(*b*) Negative when there is no agglutination in the culture-saline-serum mixture.[22] These cultures should be tested also with polyvalent H (flagellar) antiserum (III, immediately below).

(*c*) Non-specific when both mixtures agglutinate. This result requires additional testing as described in *Identification of Enterobacteriaceae* by Edwards and Ewing (1962).

22/The polyvalent O antisera do not contain agglutinins for antigens of some salmonellae. Negative somatic reactions occur with the salmonellae serotypes whose corresponding agglutinins are not contained in the antisera (such as *S. cerro*, group K(18); *S. minnesota*, group L(21); *S. alachua*, group O(35)).

III. Procedure for polyvalent H (flagellar) test

1 Dilute and adequately pretest antisera with known cultures to ensure reliability of results with unknown cultures. Follow procedures given in steps 2 to 7 immediately below.

2 To 5 ml of a 24-hour H Broth culture of the organism under test, add 5.0 ml of 0.6% formalized physiological saline solution. Let stand 1 hour before use. Formalized broth can be stored at 5–8° C for several days if necessary.

3 Place 0.02 ml of an appropriately diluted *Salmonella* polyvalent flagellar H antiserum in a small serological test tube (10 × 75 mm or 13 × 100 mm) and add 1.0 ml of the formalized broth culture (antigen).

4 If the formalized broth culture contains granular particles, pellicles, or sediment, it must also be tested substituting formalized saline in place of antiserum (saline control). Place 0.02 ml of formalized saline in the same size of serological test tube as used in the previous test (step 3 immediately above) and add 1.0 ml of the formalized broth culture.

5 Incubate the antigen-serum mixture (step 3 immediately above) and (if tested) the corresponding antigen-saline mixture (step 4 immediately above) at 50° C for 1 hour in a water bath. Observe preliminary results at 15 minute intervals and read final results at the end of 1 hour of incubation.

6 Classify the polyvalent H test as:

(*a*) Positive when there is agglutination in the culture-formalized-saline-serum mixture and no agglutination in the culture-formalized-saline mixture.

(*b*) Negative when there is no agglutination in the culture-formalized-saline-serum mixture.[23]

(*c*) Non-specific when both mixtures agglutinate. This result requires additional testing as described in *Identification of Enterobacteriaceae* by Edwards and Ewing (1962).

7 Cultures that are non-motile or cultures that are *Salmonella* polyvalent H (flagellar) negative when retested are classified according to the results of other tests as described in the manual *Identification of Enterobacteriaceae* by Edwards and Ewing (1962).

23/The polyvalent H antiserum does not contain agglutinins for antigens of some salmonellae. Negative flagellar reactions occur with the *Salmonella* serotypes whose corresponding agglutinins are not contained in the antisera (i.e., *S. simsbury* z_{27}; *S. wichita*, z_{37}; *S. chittagong*, z_{85}).

IV. Procedure for individual O group test

This test is performed to determine the O (somatic) group to which the culture belongs.

1 Dilute and adequately pretest antisera with known test cultures to ensure reliability of results with unknown cultures. Follow procedure given in steps 2 to 4 immediately below.

2 Perform O group test on culture as in II, steps 2 through 6, above, using individual group O antisera (including Vi) in place of the *Salmonella* polyvalent O antiserum. Repeat the test using each O group antiserum.

3 Cultures that are positive with Vi serum should be suspended in 1 ml of physiological saline to make a heavy suspension, then heated in boiling water for 20–30 minutes and allowed to cool. Retest heated culture suspension using O group D, C_1, and Vi sera. Vi-positive cultures which react with somatic group D serum are probably *Salmonella typhi* and Vi-positive cultures which react with somatic group C_1 serum are probably *Salmonella paratyphi* C. Heated Vi-positive cultures which do not react with any individual somatic serum but continue to react with Vi serum probably belong to the *Citrobacter* group and are not salmonellae.

4 Cultures that give a positive somatic test when tested with one of the individual O groups are recorded as positive for that group; cultures that do not react with any individual O antisera are recorded as negative.

V. Procedure for Spicer-Edwards H (flagellar) tests

These tests may be performed in place of the polyvalent H test (III above) to determine the presence or absence of H antigens.

1 Dilute and adequately pretest antisera with known cultures to ensure reliability of results with unknown culture. Follow procedures given in steps 2 and 3 immediately below.

2 Test each culture using each of the seven Spicer-Edwards H antisera. Perform this test as for the polyvalent H test (III above) using one of the seven Spicer-Edwards H antisera for each test instead of the *Salmonella* polyvalent H antiserum.

3 Positive agglutination[24] indicates the presence of an H antigen. The antigen can be identified by comparing the pattern of agglutination reac-

24/If the culture produces a positive agglutination when tested with each of the four Spicer-Edwards antisera Nos. 1, 2, 3, and 4 (a four plus pattern), then the results indicate the presence of a non-specific antigen other than a *Salmonella* antigen, or the presence of more than a single *Salmonella* H antigen, which cannot be identified with these antisera until the antigens are separated.

tions obtained with the agglutinins known to be present in each of the
seven Spicer-Edwards antisera (Table 6). The results of these reactions
are supplied also by the manufacturer in a table that lists the Spicer-
Edwards antisera and the H antigens with which each reacts.

TABLE 6

Positive reactions of Spicer-Edwards *Salmonella* H
antisera with H antigens

H antigen	Positive reaction in Spicer-Edward *Salmonella* H antisera
a	1, 2, 3
b	1, 2, 4
c	1,2
d	1, 3, 4
eh	1, 3
G complex	1, 4
i	1
k	2, 3, 4
r	2, 4
y	2
z	3, 4
z_4 complex	3
z_{10}	4
z_{29}	2, 3
enx, enz_{15}	en complex
lv, lw, lz_{13}, lz_{28}	l complex
1, 2; 1, 5; 1, 6; 1, 7; z_6	l complex

Phage typing of salmonellae

Phage typing is a desirable epidemiological aid in the identification of
salmonellae organisms for which acceptable phages are available. Production of *Salmonella* phages is under the aegis of the World Health Organization, but international distribution even of those developed to date is not
yet feasible.

Shigellae

Bacteria of the genus *Shigella* cause dysentery, and other similar but less
severe diarrhoeal illnesses, depending on the species involved. The
diseases (called shigellosis) are infectious, being spread by faecal contamination, direct or indirect; they are sometimes transmitted through
water or food contaminated by human carriers, in some countries, often
through the intervention of flies (cf. Part I, p. 12). Routine examination

of foods for shigellae has not been usual, except when shigellae may be revealed on agar media used primarily to detect *Salmonella*, for example, Salmonella-Shigella or Desoxycholate Citrate agars.

Methods specific for the isolation of shigellae from foods have received little study, but the need is becoming more widely recognized. A practice now followed of using media designed primarily for the recovery of salmonellae is not fully satisfactory (Taylor, 1965). Taylor's Xylose Lysine Desoxycholate Agar has been used effectively by some members of the Committee and is offered here with the realization that it tends to be inhibitory to *Sh. dysenteriae* and may permit false-positive reactions from *Providencia* and some *Pseudomonas* species. The non-motility of *Shigella* seems to be the best criterion for separation from these confusing species, followed by serological confirmation of *Shigella*. More research is necessary. Procedures for serological investigations of shigellae are not described here because they have not been widely used in laboratories for food microbiology. They are available from Edwards and Ewing (1962) and from the International Enterobacteriaceae Subcommittee Report (1958).

ENUMERATION OF SHIGELLAE

I. Apparatus and materials

1   Requirements for Method 1 or 2 in section on Preparation and Dilution of the Food Homogenate (p. 60).

2   Petri dishes, glass (100 × 15 mm) or plastic (90 × 15 mm).

3   1-ml bacteriological pipettes with subdivisions of 0.1 ml or less.

4   Water bath or air incubator for tempering agar, $45° \pm 1°$ C.

5   Incubator, $37° \pm 1°$ C.

6   A drying cabinet or incubator for drying the surface of agar plates. (See item 9, p. 158, for note on drying.)

7   Glass spreaders (hockey-stick-shaped glass rods, fire-polished; approximate measurements: 3.5 mm diameter, 20 cm long, bent at a 90° angle 3 cm from one end).

8   Xylose Lysine Desoxycholate Agar (XLD Agar; Medium 104, Part III), for plates.

II. Procedure

1   Pour plates of XLD Agar and dry surfaces in drying cabinet.

2   Prepare food samples by one of the two methods described in the section on Preparation and Dilution of the Food Homogenate (p. 60), using the same technique and extent of dilution.

3 Pipette 0.1 ml of food homogenate and dilutions of homogenate onto the surfaces of separate dried plates and·spread each portion with a sterile bent glass rod; in each case continue spreading until the surface appears dry. Prepare duplicate plates for each dilution.

4 Incubate plates inverted at 37° C for 24 hours.

5 Select all plates possessing between 30 and 300 colonies and count all colonies that appear uniformly red. These are considered presumptive *Shigella*. Do not count red colonies possessing black centres or amber-tinted colonies; these are likely to be species of *Salmonella* and *Arizona*.

6 Calculate the presumptive *Shigella* count from the dilutions used.

Further identification of isolates can be carried out by serological (Edwards and Ewing, 1962) and biochemical (Breed *et al.*, 1957; Edwards and Ewing, 1962) testing.

Vibrio parahaemolyticus

Vibrio parahaemolyticus is a member of a group of bacteria often referred to as the pathogenic halophiles. They are best known in Japan, where some strains are commonly associated with a dysentery-like illness following consumption of severely contaminated fish. The group has the following characteristics: Gram-negative polymorphous rods, which are motile with a single polar flagellum; facultative anaerobes; require sodium, potassium, and magnesium ions for growth; grow sparingly on most common media unless 2–3% sodium chloride is added (Blood Agar and Brain Heart Agar are exceptions in not requiring added salt); optimum temperature for growth, 30–37° C; optimum pH, 7.5–8.8; maximum pH for growth, 9.4; colony dissociation common on agar media (dissociation can be inhibited by adding a detergent of the sulphated fatty alcohol type (e.g., "Teepol") to the agar medium).

The determinative biochemical characteristics used in the identification of pathogenic halophiles are shown in Tables 7 and 8. Those listed in Table 8 are used to classify these organisms into two biotypes. Biotype 1 is called *Vibrio parahaemolyticus* and causes food-poisoning in man. Biotype 2, called *Vibrio alginolyticus*, is also believed to cause human food-poisoning under some circumstances.

Recently Kodama (1967) reported that most of the *V. parahaemolyticus* strains isolated from patients show definite haemolysis on Human

Blood Agar plates, while those from sea-water and fish samples are weakly haemolytic or non-haemolytic. This haemolytic activity decreases after serial passages of haemolytic strains in artificial media. Further study is needed to establish any relationship between haemolysis and patho- genicity.

V. parahaemolyticus has three group antigens (O, K, and H) but is classified serologically on the basis of the O and K antigens only. The O antigen is somatic, thermostable, and destroyed by ethanol and 1N HCl. The K antigen is capsular in origin and is destroyed by heating at 100° C and by 1N HCl. The presence of K antigen on the living bacteria inhibits the agglutination reaction by O antigen and antibody. To date, 10 O anti- gens and 33 K antigens have been recognized, and the combination of these O and K antigens makes the serological classification of *V. para- haemolyticus* possible (Sakazaki, 1967).

The following methods for the isolation and identification of *V. para- haemolyticus* follow those described in a publication by the Department of Health and Welfare of Japan (1963). No other country has had extensive experience with this organism. Most of the media devised for isolation of *V. parahaemolyticus* have been shown also to be excellent for isolating *Vibrio comma*.

ISOLATION OF *VIBRIO PARAHAEMOLYTICUS*

I. Apparatus and materials

1   Requirements for Method 1 or 2 in the section on Preparation and Dilution of the Food Homogenate (p. 60).

2   Petri dishes, glass (100 × 15 mm) or plastic (90 × 15 mm).

3   Incubator, 37° ± 1° C, for incubation and drying plates.

4   Water bath or air incubator for tempering agar, 45° ± 1° C.

5   Inoculating needle with a 3-mm-diameter loop, preferably of nichrome or platinum-iridium wire.

6   Peptone Salt (4%) Broth (Medium 55, Part III), 2–3 cm depth in tubes.

7   Bismuth Sulphite Salt Broth (Medium 5, Part III), 10 ml volumes in 150 × 15 mm tubes.

8   BTP Teepol Agar (Medium 13, Part III), for plates.

9   Water-Blue Alizarine-Yellow Agar (WA Agar; Medium 103, Part III), for plates.

10 Thiosulphate Citrate Bile Salts Agar (TCBS Agar; Medium 87, Part III), for plates.

11 Bismuth Sulphite Phenol-Red Agar (Medium 4, Part III), for plates.

II. Procedure

1 Prepare food samples by one of the two methods described in the section on Preparation and Dilution of the Food Homogenate (p. 60), using Peptone Salt 4% Broth as the dilution fluid instead of Phosphate-Buffered Dilution Water or Peptone Dilution Fluid. Use the same technique and extent of dilution.

2 Streak a loopful of each dilution of food homogenate on to the surfaces of separate plates of one of the four agar media listed in I above in such manner as to obtain individual colonies. Incubate plates at 37° C for 18–24 hours. All media are satisfactory; the choice will depend on the preference and experience of the individual worker.

3 Where the number of cells of *V. parahaemolyticus* in the food is likely to be small, enrich by inoculating 1 ml of undiluted food homogenate into 10 ml of Bismuth Sulphite Salt Broth and incubating the culture at 37° C for 15–24 hours.

4 Shake the enrichment culture and streak a loopful on to the surface of one of the four agar media as described in step 2 above. Incubate plates at 37° C for 18–24 hours.

5 Examine all plates and choose at least 10 well-separated colonies for identification (see procedures below). Selection of the colonies can be made on the basis of sugar fermentation; most strains of *V. parahaemolyticus* do not ferment sucrose (see Table 8).

(*a*) On BTP Teepol Agar, colonies of pathogenic halophiles that ferment sucrose are yellow (acid reaction) while those that do not are dark green (alkaline reaction). Coliforms will not grow on this medium.

(*b*) On Water-Blue Alizarine-Yellow Agar, colonies that ferment sucrose are blue (acid reaction) while those that do not are yellow (alkaline reaction).

(*c*) On Thiosulphate Citrate Bile Salts Agar, colonies that ferment sucrose are light yellow (acid reaction) while those that do not are blue (alkaline reaction). The high salt concentration and pH value of this medium suppresses most organisms other than halophiles.

(*d*) On Bismuth Sulphite Phenol-Red Agar, colonies that ferment sucrose are dark brown (acid reaction) while those that do not are without pigmentation. Coliforms will not grow on this medium.

IDENTIFICATION OF *VIBRIO PARAHAEMOLYTICUS*

I. Apparatus and materials

1   Petri dishes, glass (100 × 15 mm) or plastic (90 × 15 mm).

2   Inoculating needle, preferably with nichrome or platinum-iridium wire.

3   Water bath or air incubator for tempering agar, 45° ± 1° C.

4   Incubators, at 25–27° C and at 37° ± 1° C.

5   Refrigerator, 4° C.

6   Nutrient Salt Agar (Medium 49, Part III), slants.

7   SIM Salt Medium (Medium 75, Part III), 7–10 ml in 150 × 15 mm tubes.

8   Peptone Broths (Media 52, 54, 55, 56, 57, and 58, Part III), 2–3 cm depth in tubes.

9   Hugh-Leifson Salt Medium (Medium 31, Part III), 7–10 ml in 150 × 15 mm tubes.

10   Nutrient Salt Gelatin (Medium 50, Part III), 7–10 ml in 150 × 15 mm tubes.

11   Salts Carbohydrate Broth (Medium 69, Part III). Tubes for each of 15 carbohydrates: glucose, trehalose, maltose, mannitol, starch, cellobiose, lactose, rhamnose, xylose, adonitol, dulcitol, inositol, salicin, sucrose, and arabinose; 5 ml volumes in 150 × 15 mm tubes.

12   Triple Sugar Iron Salt Agar (Medium 89, Part III), 5–6 cm depth in tubes.

13   Jordan Salt Agar (Medium 32, Part III), 2–3 cm depth in tubes.

14   Buffered Glucose Salt Broth (Medium 15, Part III), 5 ml volumes in 150 × 15 mm tubes.

15   1% Naphthol Solution (prepare as described for Reagent 7, Part III, except add 1 g of α-naphthol per 100 ml of ethanol).

16   6% α-Naphthol Solution (prepare as described for Reagent 7, Part III, except add 6 g of α-naphthol per 100 ml of ethanol).

17   Indole Reagent (Reagent 3, Part III).

18   Sulphanilic Acid Solution (Reagent 9, Part III).

19 α-Naphthylamine Solution (Reagent 8, Part III).

20   40% aqueous solution of potassium hydroxide.

21   1% aqueous solution of *p*-aminodimethylaniline succinate.

22   Liquid paraffin.

23 Solutions (Reagent 1, Part III) and microscope slides for staining smears by Gram's method.

II. Procedure

1 Purify the cultures selected in the section on Isolation of *Vibrio parahaemolyticus*, above, II, step 5, by (*a*) streaking each on separate plates of Nutrient Salt Agar in such manner as to obtain isolated colonies, (*b*) incubating plates at 37° C for 24 hours, (*c*) transferring cells from a separated colony in each case to a slope of Nutrient Salt Agar, and (*d*) incubating the tubes in turn at 37° C for 24 hours.

2 Prepare smears on glass slides, stain by Gram's method, and examine.

3 Submit cultures appearing pure and composed of short Gram-negative rods to the following determinative tests. Use cells from 24-hour Nutrient Salt Agar slant cultures or Peptone Salt (3%) Broth cultures for the inoculum as indicated.

(*a*) *Motility*

(i) Inoculate one tube of SIM Salt Medium by holding tube in a vertical position and making a single stab with a straight needle. Incubate tube upright at 37° C for 18–24 hours.

(ii) Examine and record. Motility is indicated by a diffuse zone of growth spreading from the line of inoculation. *V. parahaemolyticus* is motile.

(*b*) *Halophilic character*

(i) Transfer cells from a 24-hour slope culture to one tube each of Peptone Broth and Peptone Salt (3%) Broth and incubate tubes at 37° C for 18 hours.

(ii) Compare growth in the two media. Cultures that grow in the Peptone Salt (3%) Broth but not in the Peptone Broth are considered halophilic.

(*c*) *Na Cl resistance*

(i) Transfer 1 loopful of a 24-hour broth culture to one tube each of Peptone Salt (7%) Broth and Peptone Salt (10%) Broth and incubate the tubes at 37° C for 24 hours.

(ii) Examine growth in tubes. *V. parahaemolyticus* grows in Peptone Salt (7%) Broth but not in Peptone Salt (10%) Broth.

(*d*) *Cytochrome oxidation*

(i) Transfer cells from a 24-hour slope culture to a tube of Peptone Salt (3%) Broth and incubate tube at 37° C for 18–24 hours.

(ii) Add 0.2 ml of 1% α-naphthol solution and 0.3 ml of 1% *p*-aminodimethylaniline succinate solution to the culture.

(iii) Shake vigorously and examine. A positive test is indicated by the development of a blue colour within 5 minutes. *V. parahaemolyticus* gives a positive cytochrome test.

(*e*) *Hugh-Leifson test*

(i) Stab inoculate two tubes of Hugh-Leifson Salt Medium with cells from a 24-hour agar slant.

(ii) Overlay one tube only with about ½ inch of liquid paraffin and incubate both tubes at 37° C for 18–24 hours.

(iii) Observe tubes for colour change for gas production. A change from the original blue colour to yellow in both tubes indicates glucose fermentation. When this change occurs only in the tube not containing paraffin, oxidative utilization of glucose is indicated. Gas production is revealed by the formation of pockets of gas in the agar. *V. parahaemolyticus* ferments glucose without gas production.

(*f*) *Indole production*

(i) Transfer a loopful of cells from a 24-hour slope culture to a tube of Peptone Salt (3%) Broth and incubate tube for 18–24 hours at 37° C.

(ii) Test for indole production by adding 0.2–0.3 ml of indole reagent to the tube. Shake tube and let stand 10 minutes before reading. A positive test is indicated by a deep red colour in the amyl alcohol layer. *V. parahaemolyticus* is indole-positive.

(*g*) *Nitrate reduction*

(i) Transfer 1 loopful of cells from a 24-hour slope culture to a tube of Peptone Salt Nitrate Broth and incubate tube at 37° C for 18–24 hours.

(ii) Add 0.5–1 ml of each of Sulphanilic Acid Solution and α-naphthylamine Solution to the culture. Shake tube and observe colour. Development of a distinct pink or red colour represents a positive nitrate reduction test (presence of nitrite). *V. parahaemolyticus* gives a positive test.

(*h*) *Gelatin liquefaction*

(i) Stab inoculate a tube of Nutrient Salt Gelatin with cells from a 24-hour agar slant and incubate culture at 37° C for 48–72 hours.

(ii) Place tube in the refrigerator at 4° C for 30 minutes and then read. A positive test is indicated by noticeable liquefaction of the gelatin. *V. parahaemolyticus* does not liquefy gelatin.

(*i*) *Voges-Proskauer test*

(i) Inoculate a tube of Buffered Glucose Salt (3%) Broth with a loopful of cells from a 24-hour slope culture and incubate tube at 25–27° C for 48 hours.

(ii) Add 1 ml of 6% α-Naphthol Solution and 0.2 ml of 40% potassium hydroxide solution to the culture and shake vigorously.

(iii) Let tube stand 2–4 hours and observe the colour. Development of a pink to crimson colour represents a positive test (production of acetylmethylcarbinol). *V. parahaemolyticus* gives a negative test.

(j) Fermentation of carbohydrates

(i) Inoculate one tube of each of the 15 fermentation broths listed in I, item 11, with a loopful of cells from a 24-hour slope culture and incubate tubes at 37° C for 18–24 hours.

(ii) Development of a yellow colour indicates a positive test fermentation. (See Table 7 for fermentation reactions for the various carbohydrates.)

TABLE 7
Biochemical characteristics of pathogenic halophiles

Halophilic character	+
Growth in Peptone salt (7%) Broth	+
Cytochrome oxidation	+
Indole production	+
Nitrate reduction	+
d-Tartrate utilization	+
Fermentation of glucose (no gas), maltose, trehalose, mannitol, and starch	+
Fermentation of cellobiose, lactose, rhamnose, xylose, adonitol, dulcitol, inositol, and salicin	–
Fermentation of sucrose	–(+)
Fermentation of arabinose	+(–)
Gelatin liquefaction	–
Hydrogen sulphide production	–
Hugh-Leifson test	Fermentation

(k) Hydrogen sulphide production

(i) Stab inoculate a tube of Triple Sugar Iron Salt Agar with cells from a 24-hour slope culture and incubate tube at 37° C for 18 hours.

(ii) Development of a black colour in the butt of the tube indicates production of H_2S and a positive test. *V. parahaemolyticus* does not produce H_2S.

(l) d-Tartrate utilization

(i) Using a straight needle stab inoculate a tube of Jordan Salt Agar with cells from a 24-hour agar slope and incubate tube at 37° C for 18 hours.

(ii) A colour change in the medium to yellow from the original red indicates a positive test. *V. parahaemolyticus* gives a positive test.

4 Based on the above tests, identify cultures with the aid of Tables 7 and 8.

TABLE 8

Classification of pathogenic halophiles

	V. parahaemolyticus	*V. alginolyticus*
NaCl resistance		
7%	+	+
10%	−	+
Acetylmethylcarbinol production	−	+
Sucrose fermentation	−(+)	+(−)
Arabinose fermentation	+(−)	−(+)

Staphylococci

Staphylococci are enumerated in foods for three main reasons: (1) to examine food suspected to have caused food-poisoning, for which the presence of large numbers of coagulase-positive staphylococci would be suggestive; (2) to demonstrate a risk that the test food may be capable of inducing staphylococcal food-poisoning if stored under conditions conducive to toxin formation or if used as an ingredient in some other foods; and (3) to establish the occurrence of objectionable post-processing contamination. For some processed foods, *Staphylococcus aureus* is a good indicator of unsatisfactory sanitation within the food factory.

Most foods involved in staphylococcal food-poisoning contain several million viable staphylococci per gram. As few as 50,000 per g have been implicated in poisonings from relatively fresh food. However, the investigator must appreciate the survival of enterotoxins even after heat processes which kill the staphylococcal cells, or after prolonged storage, during which a marked reduction in numbers of staphylococci can occur.

Detection of *Staphylococcus aureus* alone does not constitute satisfactory proof of involvement of the source food in food-poisoning. Tests for detection of one or more of the causative enterotoxins are desirable. Methods for detection of these enterotoxins are not elaborated here because the necessary diagnostic antitoxins are not available commercially and research to establish the full number of such toxins is not complete.

There are a number of methods available for the enumeration of staphylococci. They differ mainly in the nature of the selective agents used, chief of which are high concentrations of sodium chloride, sodium tellurite, lithium chloride, and sodium azide (Crisley *et al.*, 1965). Rather recently, media containing egg yolk together with one or more of these selective agents have been introduced. In egg-yolk media *Staph. aureus* utilizes the egg yolk lipoprotein, lipovitellenin (Shaw and Wilson, 1963; Tirunarayanan and Lundbeck, 1967), and this results in the formation of cleared areas under and around colonies. Another identifying feature of the "egg-yolk reaction" is for formation of a white precipitate in the cleared or partially cleared areas, due to the formation of calcium and magnesium salts of liberated fatty acids (Tirunarayanan and Lundbeck, 1967).

Of the many methods available for the determination of staphylococci, five are described here. The Baird-Parker medium is now widely used, particularly in Europe (Baird-Parker, 1962b). At the time of the Committee's discussion the quality of commercially available Baird-Parker medium was under some doubt. However, the dehydrated product is now satisfactory. Results of a comparative study (Crisley et al., 1965), had not indicated a preferred rank for this medium, but the newer satisfactory product was not used in that study.

Three other methods using the egg-yolk reaction are described. Each is employed with evident satisfaction: one in Sweden (Lundbeck and Tirunarayanan, 1966), involving Egg-Yolk Azide Agar; one in the Soviet Union (Meshalova and Mikhailova, 1964), employing Salt Egg-Yolk Agar; and one in the United States (Crisley et al., 1964, 1965), using Tellurite Polymyxin Egg-Yolk Agar.

A method based on a tellurite-glycine medium (Vogel and Johnson, 1960) is also described. It has recently undergone extensive comparative (Baer et al., 1966) and collaborative (Baer, 1966) testing in the United States. The United States Food and Drug Administration has submitted the method to the Association of Official Analytical Chemists for adoption.

Pending interlaboratory trials, the Committee has decided to recommend all five methods.

ENUMERATION OF COAGULASE-POSITIVE STAPHYLOCOCCI

Method 1[25]

I. Apparatus and materials

1 Petri plates, glass (100 × 15 mm) or plastic (90 × 15 mm).
2 1-ml bacteriological pipettes with subdivisions of 0.1 ml or less.
3 Water bath or air incubator for tempering agar, 45° ± 1° C.
4 Incubator, 37° ± 1° C.

25/The method and medium are those described by Baird-Parker (1962a, b). This method is used widely, particularly in the United Kingdom.

5   A drying cabinet or incubator for drying the surfaces of agar plates.

6   Glass spreaders (hockey-stick-shaped glass rods, fire-polished). Approximate measurements: 3.5 mm diameter, 20 cm long, bent at a 90° angle 3 cm from one end.

7   Baird-Parker Agar (Medium 1, Part III), for plates.

II. Procedure

1   Prepare food samples by one of the two methods described in the section on Preparation and Dilution of the Food Homogenate (p. 60), using the same technique and extent of dilution.

2   Pour plates of Baird-Parker Agar (15 ml in each) and dry surfaces in drying cabinet or incubator (see item 9, p. 158, for note on drying plates).

3   Pipette 0.1 ml of homogenate and dilutions of homogenate onto the surfaces of separate plates and spread each portion with a sterile bent glass rod; in each case continue spreading until surface appears dry. Duplicate plates should be prepared for each dilution.

4   Incubate plates inverted at 37° ± 1° C for 24 and 48 hours.

5   After 24 hours of incubation, select plates possessing 30–300 separate colonies and count all colonies which are black and shiny with narrow white margins, and surrounded by clear zones extending into the opaque medium. These are colonies of *Staphylococcus aureus*.

6   Mark the position of these colonies and incubate plates for a further 24 hours.

7   Count all colonies with the above appearance that developed during the extended period of incubation and submit these, or a significant number of them (square root of count) if the count is high, to a coagulase test procedure (see section on Testing for Coagulase Production below, following Method 5).

8   With some batches of egg yolk, colonies of a few strains of *Staph. aureus* may be surrounded by an opaque zone at 24 hours, and a larger number of strains may show this appearance after 48 hours. Count these after 48 hours and submit them, or a suitable number of them (square root of the count) if there are many, to a coagulase test. This will distinguish these cultures (coagulase-positive) from *Staph. epidermidis* (coagulase-negative) which may give a similar appearance.

9   Total the colonies which produced clear zones after 24 hours of incubation, and those in steps 7 and 8 above which were coagulase-positive, and calculate from the dilutions used the total number of coagulase-positive staphylococci.

Note Over 90% of all coagulase-positive staphylococci show the characteristic black colony surrounded by a clear zone after incubation for 24–26 hours at 37° C; it is important to read results at this time and *not* at times shorter than 24 hours. A further 5–7% of coagulase-positive staphylococci show this characteristic clearing, often with an inner opaque zone, after 48 hours, at which time coagulase-negative staphylococci may also show this appearance; it is for this reason that the coagulase test should be made on suspect colonies appearing at 48 hours.

Method 2[26]

I. Apparatus and materials

1 Requirements for Method 1 above, items 1 through 5.

2 Inoculating needle with nichrome or platinum-iridium wire.

3 Vogel-Johnson Agar (Medium 102, Part III), for plates.

4 Trypticase Soy Broth with 10% Sodium Chloride (Medium 96, Part III), 5–10 ml volumes in tubes.

II. Procedure

1 Prepare food samples by one of the two procedures in the section on Preparation and Dilution of the Food Homogenate (p. 60), using the same technique and extent of dilution. In cases where food samples contain very high numbers of staphylococci, dilutions higher than 10^{-5} may be needed, since this method depends on at least one dilution giving negative results.

2 Inoculate single tubes of Trypticase Soy Broth (10% sodium chloride) with 1 ml aliquots of decimal dilutions of the food homogenate. The maximum dilution of sample must be sufficiently great to yield negative results for at least one dilution.

3 Incubate the tubes at 35–37° C for 48 hours.

4 Using a 3 mm loop, transfer 1 loopful from each inoculated tube to a separate plate of Vogel-Johnson Agar and streak in such a manner as to give isolated colonies.

5 Incubate plates at 35–37° C for 48 ± 2 hours.

6 Select at least one of each visibly different colony type, which has reduced tellurite, from all sample dilutions tested, and test these for

26/Recommended for enumeration of coagulase-positive staphylococci in foods by the U.S. Food and Drug Administration (Baer, 1966).

coagulase production (for procedure, see section on Testing for Coagulase Production below, following Method 5).

7 From highest dilution containing coagulase-positive staphylococci, estimate the number in the original specimen.

Method 3[27]

I. Apparatus and materials

1 Requirements for Method 1, above, items 1 through 6.

2 Egg-Yolk Azide Agar (Medium 20, Part III), for plates.

II. Procedure

1 Prepare food samples by one of the two methods described in the section on Preparation and Dilution of the Food Homogenate (p. 60), using same technique and extent of dilution.

2 Pour plates of Egg-Yolk Azide Agar and dry surfaces in drying cabinet or incubator (see item 9, p. 158, for note on drying plates).

3 Pipette 0.2 ml of homogenate and dilutions of homogenate onto the surfaces of separate plates and spread each portion with a sterile bent glass rod; in each case continue spreading until surface appears dry. Duplicate plates should be used for each dilution.

4 Incubate plates inverted at 37° ± 1° C for 24 and 48 hours. After 24 hours, clear areas (haloes) containing little or no granular precipitate surround colonies of staphylococci. The surface of the halo has an oily appearance due to Newton ring phenomena following the liberation of free fatty acid by a lipase esterase action of the staphylococci. After 48 hours the haloes are larger and contain a white granular precipitate (calcium salts of fatty acids). If the colony is yellow, owing to pigmentation, the whole picture resembles a fried egg.

5 Count all colonies having the above characteristics, on plates possessing between 30 and 300 colonies.

6 Test these colonies, or a significant number of them (square root of the count) if the count is high, for coagulase production (for procedure, see section on Testing for Coagulase Production below, following Method 5).

7 From the proportion which is coagulase-positive and the dilution, calculate the total count in the specimen and report as coagulase-positive staphylococci.

27/Used in a number of laboratories in Europe, particularly in the Scandinavian countries (Lundbeck and Tirunarayanan, 1966).

Notes The correlation between the lysis precipitate and coagulase production is currently being studied (H. Lundbeck, personal communication). Although most of the coagulase-positive staphylococci produce the lipase-esterase enzymes, a few coagulase-negative strains also demonstrate the lytic-precipitation phenomenon.

Sodium azide at the concentrations used does not inhibit staphylococci but prevents growth of all Gram-negative organisms except some strains of *Proteus* and *Pseudomonas*.

Method 4[28]

This method includes procedures for direct plating and enrichment culturing.

I. Apparatus and materials

1 Requirements for Method 1 above, items 1 through 6.

2 Milk Salt Agar (Medium 43, Part III), for plates.

3 Salt Egg-Yolk Agar (Medium 70, Part III), for plates.

4 Blood Agar (Medium 6, Part III), for plates.

5 Salt Broth 6.5% (Medium 67, Part III), 7–10 ml volumes in tubes.

6 Salt Broth 10% (Medium 68, Part III), 7–10 ml volumes in tubes.

7 Glucose Broth (Medium 27, Part III), 7–10 ml volumes in tubes.

8 Nutrient Agar (Medium 46, Part III), slants in tubes.

9 Materials for staining smears by Gram's method (Reagent 1, Part III).

II. Procedures

A. DIRECT PLATING

1 Prepare food samples by one of the two procedures described in the section on Preparation and Dilution of the Food Homogenate (p. 60), using the same technique and extent of dilution.

2 When estimating the total number of cells, pipette 0.1 ml of food homogenate and dilutions of homogenate onto the surface of separate, previously prepared and dried plates of Milk Salt Agar. Spread each portion with a sterile bent glass rod or inoculating needle over entire surface of plate, continuing the spreading action until surface appears dry.

28/This method is used throughout the Soviet Union (Nefedjeva, 1964) for examination of foods for coagulase-positive staphylococci and in epidemiological studies.

3 Incubate plates at 37° ± 1° C for 24 and 48 hours.

4 After 24 hours of incubation, select plates containing 30–300 separated colonies and count all colonies which are surrounded by a cleared zone.

5 Mark the position of these colonies and incubate selected plates for a further 24 hours.

6 Count typical colonies that developed during the extended incubation period and add the total to the 24-hour count.[29]

7 Select a significant number of typical colonies (square root of total if count is high) on the plates counted and transfer these to slopes of Nutrient Agar. Incubate slopes at 37° ± 1° C for 24 hours.

8 Prepare smears of slope cultures and stain them by Gram's method.

9 If cultures are Gram-positive and appear pure, submit them to the following tests:

(*a*) *Coagulase test*   Follow procedure described in the section on Testing for Coagulase Production below, following Method 5.

(*b*) *Lipase reaction*   Streak the surface of a previously prepared dried plate of Salt Egg-Yolk Agar with a loopful of 24-hour cells and incubate plate at 37° ± 1° C for 24 hours. Staphylococci are lecithinase-positive and will clarify the medium around colonies.

(*c*) *Haemolysis test*   Streak the surface of a previously prepared dried plate of Blood Agar with a loopful of 24-hour cells and incubate plate at 37° ± 1° C for 24 hours. Enterotoxin-producing staphylococci are usually haemolytic.

10 From the highest dilution containing coagulase-positive staphylococci, calculate the total number of staphylococci in the original food specimen. These strains are usually lipase-positive and have haemolytic properties. Omit this step if an estimation of the total count is not required.

B. ENRICHMENT CULTURING

If it is necessary to determine the presence or absence of coagulase-positive staphylococci in foods likely to contain only a small number of cells, the following enrichment procedure may be carried out simultaneously with the direct plating method described above in A of this method.

29/If the determination of the total number of cells is not required, steps 4, 5, 6, and 10 can be omitted.

1 Pipette 1 ml of the original homogenate (1:10 dilution) into a tube of each of the following three enrichment broths: Salt Broth 6.5%, Salt Broth 10%, and Glucose Broth. Incubate tubes at 37° ± 1° C for 18–24 hours. When examining foods containing a very high salt content (e.g., salted fish), Salt Broth 10% can be omitted. Similarly, when examining sweet foods such as frostings and ice cream, Glucose Broth can be omitted.

2 Streak a loopful of each broth culture onto the surface of a separate, previously prepared and dried plate of Milk Salt Agar. Repeat using Blood Agar. Incubate plates at 37° ± 1° C for 24 hours.

3 Select several suspect colonies from each plate and transfer cells to Nutrient Agar slopes. Incubate slopes at 37° ± 1° C for 24 hours.

4 Submit slope cultures to tests outlined in A, steps 8 and 9, omitting the haemolysis test for cultures isolated from Blood Agar plates.

Method 5[30]

I. Apparatus and materials

1 Requirements for Method 1 above, items 1 through 6.

2 Tellurite Polymyxin Egg-Yolk Agar (TPEY Agar; Medium 80, Part III), for plates.

II. Procedure

1 Pour plates of TPEY Agar and dry surfaces (see item 9, p. 158, for note on drying techniques).

2 Prepare food samples by one of the two procedures described under Preparation and Dilution of the Food Homogenate (p. 60), using the same technique and extent of dilution.

3 Pipette 0.1 ml of food homogenate and dilutions of homogenate onto the surfaces of separate plates and spread each portion with an individual glass spreader; in each case continue spreading until surface appears dry again.

4 Incubate plates inverted at 35–37° C for 24 and 48 hours and examine. After 24 hours staphylococci colonies are about 1.0–1.5 mm in diameter, appear jet black or dark grey in colour, and exhibit one or more of three types of egg-yolk reaction: (i) a discrete zone of precipi-

30/Used in a number of laboratories in the United States (Crisley *et al.*, 1964, 1965).

tated egg yolk around and beneath the colony, (ii) a clear zone or halo often occurring together with a zone of egg-yolk precipitation beneath the colony, and (iii) absence of either a precipitation zone or halo but visible precipitation beneath the colony. Egg-yolk reactions are generally better after 48 hours of incubation.

5 Count all jet black colonies giving typical egg-yolk precipitation reactions on plates possessing between 30 and 300 colonies.

6 Test colonies, or a significant number of them (square root of the count) if the count is high, for coagulase production (see procedure immediately below).

7 From the proportion of selected colonies proving to be coagulase-positive, calculate the number of coagulase-positive staphylococci per gram of the original food sample.

TESTING FOR COAGULASE PRODUCTION

I. Apparatus and materials

1 1-ml bacteriological pipettes with 0.1 ml gradations or smaller.

2 Inoculating needle with nichrome or platinum-iridium wire.

3 Incubator, 35–37° C.

4 Brain Heart Infusion Broth (Medium 8, Part III), 5 ml volumes in tubes.

5 Rabbit Plasma (Medium 63, Part III), 0.3 ml in small tubes, approximately 10 × 75 mm in size.

II. Procedure

1 Subculture selected colonies (see section on Enumeration of Coagulase-Positive Staphylococci above, Methods 1 through 5) in Brain Heart Infusion Broth and incubate 20–24 hours at 35–37° C.

2 Add 0.1 ml of resulting cultures to 0.3 ml of Rabbit Plasma in small tubes and incubate at 35–37° C.

3 Examine tube for clotting after 4 hours and, if not positive, again after 24 hours. The formation of a distinct clot is evidence of coagulase activity. See Figure 1 for types of coagulase test reactions; note that only a 2+ reaction or greater is considered positive evidence of coagulase production.

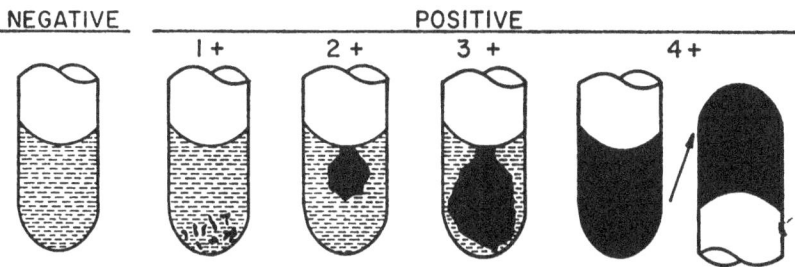

NEGATIVE   POSITIVE

1 +   2 +   3 +   4 +

FIGURE 1 Types of coagulase test reactions (from Turner and Schwartz (1958), published by the C. V. Mosby Company, St. Louis, Missouri). Note that only a 2+ or greater reaction is considered positive evidence of coagulase production. (Negative: no evidence of fibrin formation. 1+ positive: small unorganized clots. 2+ positive: small organized clots. 3+ positive: large organized clot. 4+ positive: entire content of tube coagulates and is not displaced when tube is inverted.)

Clostridium botulinum

Clostridium botulinum produces one of the most highly toxic substances known and causes the most dangerous form of food-poisoning. The organism is not normally enumerated from foods, but efforts are generally focused on detecting the presence of one of the specific types of botulinus toxins in the suspected food. The basis for choosing the most appropriate test procedure is described in Part I (p. 19). Since the organisms very closely resemble other common non-toxinogenic clostridia, immunological detection of the specific toxin remains the essential procedure. However, direct microscopic examination of the food sample should be carried out. Typical Gram-positive bacilli with subterminal oval spores may (or may not) be present.

SCREENING TEST FOR DETECTION OF *CLOSTRIDIUM BOTULINUM* TOXIN

I. Apparatus and materials

1 Mechanical food blendor capable of homogenizing sample volumes as small as 4 ml and as large as 20 ml.

2 Centrifuge, clinical, tube type, capable of attaining a speed of at least 1000 rpm.

3 Suitable glass containers for incubating and centrifuging food specimens.

4 Sterile saline (0.85% aqueous solution of sodium chloride).

5 Pipettes, 2- and 5-ml capacities, and pipette suction bulb.

6 Mice.

7 Hypodermic needle for injection of mice.

8 Polyvalent and monovalent type-specific antitoxins (*C. botulinum* types A–F).

9 Incubator, 30° ± 1° C.

II. Procedure

Step 1 below should not be used in the preliminary examination of foods involved in outbreaks of botulism, because the incubating procedure may inactivate existing toxin. Suspected or implicated food specimens should be submitted to the screening procedure starting with step 2. Very often these foods contain a readily identifiable toxin, and screening may then permit antitoxin therapy to be pursued sooner and with greater likelihood of success.

1 Incubate 2–10 g of food material believed suitable for outgrowth of spores (e.g., fish and fish products, canned foods, cooked meat dishes) in test tube or glass sample jar at 30° C for 7–10 days.

2 Homogenize the incubated sample with an equal weight of sterile saline, using a mechanical blendor.

3 Centrifuge the homogenate at high speed for 1 hour, preferably in a refrigerator or cold room.

4 Subject supernatant extract to toxicity tests by intraperitoneal injection of mice:

(*a*) Pipette 2 ml of supernatant to an empty test tube and place tube in boiling water bath for 10 minutes.

(*b*) Inject one mouse with 0.5 ml of the boiled supernatant. This is the control test for a heat-labile toxin.

(*c*) For each available polyvalent or monovalent anti-toxin inject two mice with 0.5 ml of unheated supernatant.

(*d*) Protect one of the two mice with the appropriate antitoxin, using a suitably chosen volume and dilutions to provide 1000 mouse-protective doses of antitoxin. This strength of antitoxin assures neutralization of toxin in the event an exceptionally large amount is present in the food.

5 Observe test animals for up to 4 days. The presence of botulinus toxin is indicated by the occurrence, in mice unprotected by specific antitoxin, of a progressive motor paralysis. When large amounts of toxin are present, the first signs may be apparent within 4–5 hours in the form of an indrawn or scaphoid abdomen with "bellows" respiration, followed by dragging of the hind legs, and increasing weakness. With type E toxin, the period before the onset of symptoms tends to be shorter, and death generally occurs within 24 hours. The incubation period with other types of botulinus toxins may be prolonged to 8–10 hours and death may be delayed for a few days. Animals receiving toxin neutralized by botulinus antitoxin of a homologous type will survive, thus permitting the toxin type to be identified.

DIRECT CULTURAL PROCEDURE FOR DETECTION OF *CLOSTRIDIUM BOTULINUM* IN SUSPECT FOODS

I. Apparatus and materials

1 Requirements listed above under Screening Test for Detection of *C. botulinum* Toxin, I, items 1 through 8.

2 Cooked Meat Medium (Medium 17, Part III), or Trypticase Glucose Medium (Medium 90, Part III), 4–5 cm depth in 150 × 15 mm tubes.

3 Incubator, 30° ± 1° C.

II. Procedure

1 Place about 5 g of homogenized food sample into each of three tubes of freshly boiled and cooled medium (Cooked Meat Medium or Trypticase Glucose Medium).

2 Heat one of the tubes to 60° C for 15 minutes and another to 80° C for 30 minutes in water baths. Immerse the tubes to a depth greater than the level of material in them. Agitate tubes during the first 2 minutes of heating. Leave the third tube unheated.

3 Incubate all three tubes at 30° C until vigorous growth is visible, as evidenced by gas production, turbidity, and in some instances partial hydrolysis of the meat particles.[31] As a rule toxin is present in the highest concentration after the period of active growth and gas production. Generally no more than 4–5 days of incubation are necessary for optimal toxin production, but to cover the possibility of delayed germination of botulinus spores, tubes should be held and re-examined after 10–14 days before they are discarded.

4 Test the supernatant, or sterile filtrate (sinter glass or membrane filter) of the culture, for botulinus toxins by intraperitoneal injection, as described in the Screening Test for Detection of *C. botulinum* Toxin, II, steps

31/If the isolation of pure cultures of the causative organism is to be attempted, plates should be made at this time according to directions given in the section on Isolation of Causative Organisms in Implicated Foods below.

4 and 5 above, into mice, both unprotected and protected with mono-valent type-specific antitoxins. Development of typical botulinic symptoms that can be prevented by type-specific antitoxin confirms the presence of *C. botulinum.*

ISOLATION OF CAUSATIVE ORGANISMS IN IMPLICATED FOODS

Attempts to isolate the organism from implicated foodstuffs may be un-successful. Experience in isolating *C. botulinum* is an important factor in successful isolations of this organism from highly contaminated samples. In experienced hands, plating from Cooked Meat Medium onto both Blood Agar and Brain Heart Agar plates has proved satisfactory (Dolman, 1957, 1964). One of the following two methods may also be tried:

1 The sample or suspect culture is incubated in an enrichment medium, followed by treatment with alcohol to kill vegetative cells (Johnson *et al.*, 1966) and subsequent streaking on Blood and Egg-Yolk Agar (McClung and Toabe, 1947).

2 Enrichment in Sulphite Iron Broth with added polymyxin (Mossel *et al.*, 1956; Gibbs and Freame, 1965; Narayan, 1967). This is followed by alcohol treatment and incubation as described under 1 immediately above.

I. Apparatus and materials

1   Requirements listed above under Screening Test for Detection of *C. botulinum* Toxin, I, items 6 through 8.

2   Inoculating needle, preferably with nichrome or platinum-iridium wire.

3   Petri plates, glass, with absorbent covers such as (*a*) glazed porcelain covers or (*b*) Brewer aluminium Petri dish covers, with absorbent discs.

4   Highly efficient anaerobic incubation facilities (see p. 128, I, step 4 and p. 133, I, step 8).

5   Cooked Meat Medium (Medium 17, Part III) or Trypticase Glucose Medium (Medium 90, Part III), 4–5 mm depth in 150 × 15 mm tubes.

6   Blood Agar (Medium 6, Part III) and Brain Heart Infusion Agar (Medium 7, Part III); or Reinforced Clostridial Agar (RCM; Medium 64, Part III), for plates.

II. Procedure

1   Streak a loopful of each of the three cultures described above under Direct Cultural Procedure for Detection of *C. botulinum* in Suspect Foods,

II, step 3, onto the surface of separate plates of Blood Agar and Brain Heart Infusion Agar in such a manner as to obtain separated colonies. To prevent spreading growth as much as possible, dry the surfaces of the agar plates thoroughly before use (see item 9, p. 158, for note on drying plates). Reinforced Clostridial Agar can also be used in place of Blood Agar and Brain Heart Infusion Agar.

2 Incubate plates anaerobically at 30° C. Colonies usually appear within 1–2 days. For characteristic appearances of C. *botulinum* colonies on Brain Heart and Blood Agar media, see Dolman (1964).

3 Select isolated colonies from the plates and inoculate cells from each into tubes of freshly boiled and cooled Cooked Meat Medium or Trypticase Glucose Broth.

4 Incubate anaerobically at 30° C for 4–5 days and confirm the type of organism serologically as described above, under Screening Test for Detection of C. *botulinum* Toxin, II, steps 3 through 5. (If Cooked Meat Medium is used, toxin production can occur without anaerobic incubation.)

Clostridium perfringens

The enumeration of C. *perfringens* in foods is usually carried out to investigate the suspected or potential involvement of this organism in food-poisoning. Although foods responsible for C. *perfringens* food-poisoning are expected to have a high count (10^6 cells or more per gram) if examined immediately after the outbreak, small numbers may indicate (1) a rapid drop in the count of vegetative cells due to adverse conditions such as a low pH or freezing; (2) uneven distribution of the organism throughout the food, such as in roasts of meat or poultry; or (3) uneven temperature of reheating.

Of the five serotypes A to E, differentiated by soluble antigens, two types, A and C, are known to cause food-borne disease in man. All strains of type A, when present in foods in large numbers, should be suspect, since many diverse strains of type A have been shown to cause the characteristic illness in man.

The methods selected are essentially based on (1) haemolytic reactions on a selective Horse-Blood Agar containing neomycin sulphate, and (2) the formation of black colonies on sulphite agar media highly selective for clostridia by virtue of their content of inhibitory substances (polymyxin, sulphadiazine, and neomycin). The former principle is widely applied in the United Kingdom and the latter in Continental Europe, the United States, and Canada. In two of the methods described, the food homogenate, or dilutions of it, is streaked on the surface of agar plates, while in the third the inoculum and the agar medium are mixed and incubated in non-permeable plastic pouches. All three procedures have been effective in the investigation of food-poisoning. Suspect colonies provided by each procedure should receive confirmatory tests. Where enumeration of small numbers of *C. perfringens* is desired, an MPN procedure based on the Blood Agar reaction may be used. For a discussion of the basis for the different methods for the isolation of *C. perfringens* the reader is referred to Part I, p. 21.

ENUMERATION, ISOLATION, AND IDENTIFICATION OF *CLOSTRIDIUM PERFRINGENS*

Method 1[32]

This method includes procedures for the plate count determination, culture identification (presumptive *C. perfringens* serological typing, and carbohydrate fermentations), enrichment culturing, and determination of the most probable number.

I. Apparatus and materials

Note that the equipment and materials required for all the procedures described under Method 1 are included under this heading.

1   Requirements listed in Method 1 of the section on Preparation and Dilution of the Food Homogenate (p. 61), items 1 through 5.

2   Incubator, 35–37° C.

3   Petri dishes, glass (100 × 15 mm) or plastic (90 × 15 mm).

4   Anaerobic jar with equipment and materials (pyrogallol or evacuation pump and nitrogen, hydrogen, or a mixture of 10% carbon dioxide and 90% nitrogen) for establishing anaerobic conditions in the jar.

5   Inoculating needle, preferably with nichrome or platinum-iridium wire.

6   Materials for staining smears by Gram's method (Reagent 1, Part III).

7   Screw-capped sample jars, 8 oz approximately.

8   Dropping pipette (see Method 3 of the section on Enumeration of

32/This method is used in numerous laboratories in the United Kingdom and elsewhere (Hobbs *et al.*, 1953; Hobbs, 1965).

Mesophilic Aerobes above, p. 68, for details on the construction of this pipette).

9 Neomycin sulphate, 1% aqueous solution.

10 Ringer Solution (Medium 65, Part III), ¼ strength.

11 Physiological saline solution (0.85% w/v aqueous solution of sodium chloride).

12 Physiological saline solution containing 0.4% phenol.

13 Horse-Blood Agar (Medium 30, Part III), for plates.

14 Lactose Egg-Yolk Milk Agar (Medium 35, Part III), for plates.

15 Cooked Meat Medium (Medium 17, Part III), 4 cm depth in 150 × 15 mm tubes.

16 Antisera for 17 strains of *Clostridium perfringens* (type A) prepared as described by Hobbs *et al.* (1953).

17 *C. perfringens* antitoxin A or polyvalent *C. perfringens* therapeutic antitoxin.

18 Phenol-Red Carbohydrate Broths (Medium 60, Part III), 5 ml volumes in culture tubes for the following carbohydrates: lactose, glucose, sucrose, maltose, mannitol, dulcitol, dextrin, and starch.

II. Procedure for anaerobic and aerobic plate counts

1 Weigh food into a sterile Waring blendor jar. The amount used should be sufficient to cover the cutting blades and should be representative of the specimen being examined. Thaw small frozen specimens, or aseptically sawn or drilled pieces from large specimens, overnight at approximately 5–10° C before sampling. Operate the blendor for 1–2 minutes according to the specifications given in either Method 1 or 2 of the section on Preparation and Dilution of the Food Homogenate (p. 60). If necessary for proper blending, add sufficient quarter-strength Ringer solution to give a 1/5 or 1/10 dilution. For most suspect foods this step is not necessary. Examine powdered foods directly as described in step 2 immediately below.

2 Place 10 g of the macerated food, or 10 g of a powdered product, into a sterile screw-capped glass jar (8-oz size). Add 5 ml of sterile quarter-strength Ringer solution from a 100 ml quantity, and emulsify with the aid of a small spatula if necessary.

3 Inoculate each of two dried plates of Horse-Blood Agar with one loopful (3 mm) of the emulsified sample in such manner as to give isolated colonies. (*Note* Neomycin sulphate, 3 drops (approx. 0.06 ml) of a 1% solution, spread over the dried surface of Blood Agar plates immediately before inoculation reduces the growth of facultative anaerobic organisms.)

4   Incubate the plates at 35–37° C for 48 hours, one plate aerobically and the other anaerobically.

5   Add the remaining 95 ml of Ringer Solution to the emulsified sample in 2 above, making a 1/10 dilution. Prepare additional dilutions if required.

6   Using the drop plate method (see Method 3 of the section on Enumeration of Mesophilic Aerobes, p. 68), inoculate duplicate plates of Horse-Blood Agar from each dilution.

7   Incubate one set of plates aerobically and the other anaerobically at 35–37° C for 48 hours.

8   Count the colonies on plates from step 4 above, which give between 30 and 300 colonies. For the drop plates, count non-confluent drop areas which give 20 colonies or less. Record the average anaerobic and aerobic count. The aerobic count indicates the extent and to some degree the nature of the background flora, and enables the calculation of the concentration of anaerobes relative to that of aerobes in the food specimens.

III. Procedure for determination of presumptive *C. perfringens*

1   From the plates used in the determination of the anaerobic count select typical, well-isolated colonies for identification and note haemolysis.

2   Prepare smears from these colonies and stain by Gram's method.

3   Transfer into tubes of Cooked Meat Medium or onto Horse-Blood Agar slopes cells of colonies showing short, thick Gram-positive rods, occurring singly, in pairs, or less frequently in short chains. Incubate anaerobically at 35–37° C for 18–24 hours.

4   To verify purity, streak cells of each culture onto separate plates of Horse-Blood Agar in such manner as to obtain separate colonies. Incubate plates anaerobically for 24 hours and use cells from a typical colony in each case to conduct serology tests (see section IV below).

5   Also transfer cells from cultures in step 3, immediately above, onto dried plates of Lactose Egg-Yolk Milk Agar for Nagler reaction and lactose fermentation.

(*a*) Mark each plate in half on the underside with a wax pencil.

(*b*) On one half spread a few drops of *C. perfringens* antitoxin A of British Pharmacopoeial strength or a few drops of polyvalent *C. perfringens* therapeutic antitoxin and allow to dry.

(*c*) Streak a line of culture across the two halves of the plate starting from the side without antitoxin. Space the streaks so that four different isolates can be tested on the same plate.

(*d*) Incubate the plates anaerobically for 18–24 hours.

(*e*) The test is positive for *C. perfringens* if a dense white area appears around the streak growing on the side without antitoxin, and is absent around the streak on the side with antitoxin. A positive test denotes the production of α-toxin (lecithinase C). The medium surrounding the entire length of the streak will be pink to red in colour as a result of lactose fermentation.

(*f*) Leave plates exposed to air for 1–2 hours and examine streaks for colour. *C. perfringens* cultures, originally cream-coloured, will become red because of the fermentation of lactose. At the same time the pink coloration of the medium will tend to fade. *C. bifermentans* gives a positive Nagler reaction, but is easily distinguished from *C. perfringens* because it produces an abundance of spores and does not ferment lactose.

6 From the percentage of isolates proving to be *C. perfringens*, calculate the number of cells of *C. perfringens* in the food specimen from the anaerobic plate count (II, step 8, above).

IV. Procedure for serological typing (Hobbs *et al.*, 1953)

1 Suspend cells from the 24-hour Horse Blood Agar plate culture (see III, step 4, above) in a drop of physiological saline solution on a glass slide.

2 Mix a loopful of pooled antisera with the suspension and examine for agglutination.

3 When agglutination occurs, repeat steps 1 and 2 for this culture using the individual antisera diluted 1:5 with 0.4% phenol saline. Homologous strains will agglutinate rapidly.

V. Procedure for carbohydrate fermentation tests

As a rule sufficient evidence of the presence of *C. perfringens* is obtained from the appearance of the colony, the Nagler reaction, and serology, but it may be desirable to carry out further tests of identification by carbohydrate fermentation.

1 Inoculate cells from a 24-hour culture (in Cooked Meat Medium) of each isolate into a separate tube of Phenol-Red Carbohydrate Broth for each of the following: lactose, glucose, sucrose, maltose, mannitol, dulcitol, dextrin, and starch.

2 Incubate tubes anaerobically at 35–37° C for 48 hours.

3   *C. perfringens* produces acid and gas from all the carbohydrates listed
except dulcitol.

VI. Procedure for enrichment culturing

Where it is necessary only to determine the presence or absence of *C. perfringens* in a food and where the number of cells is likely to be so low as
to be missed by the direct plating procedure above, the following enrichment procedure is recommended.

1   Add 10 g of the macerated or powdered specimen (II, step 1, above)
to each of two preparations of 100 ml of Cooked Meat Medium in 8-oz
screw-capped sample jars. Mix well by shaking. (*Note*   Neomycin, 1.2 ml
of a 1% solution added to each 100 ml of Cooked Meat Medium, is useful
for the selection of *C. perfringens* from mixed (impure) cultures.)

2   Heat one preparation to 80° C and hold at this temperature for 10
minutes. Do not heat the other.

3   Incubate both preparations at 35–37° C for 18–24 hours.

4   Prepare duplicate smear plates on Horse-Blood Agar for each of the
resulting cultures.

5   Incubate one set of plates aerobically and one anaerobically for 24
hours at 35–37° C.

6   Identify suspicious colonies as described in III, IV, and V, above.

VII. Procedure for determination of the most probable number

1   Using food suspensions treated as described in II, step 1, above, as
source material, prepare decimal dilutions (10^{-1}, 10^{-2}, 10^{-3}, 10^{-4}, 10^{-5})
using 9- or 90-ml blanks of Ringer Solution. Shake each dilution vigor-
ously 25 times in a 1-foot arc.

2   Transfer 1 or 10 ml of each dilution to each of three or five tubes or
bottles (depending on choice of three-tube or five-tube MPN series) of
Cooked Meat Medium.

3   Incubate tubes anaerobically at 35–37° C for 24 hours.

4   From all tubes showing growth, prepare smear plates on Horse-Blood
Agar.

5   Incubate plates anaerobically at 35–37° C for 48 hours.

6   Identify typical *C. perfringens* colonies as described in III, IV, and V
above.

7   From the number of tubes shown to contain *C. perfringens* from each
dilution, determine the MPN from either Table 2 or Table 4 as directed

in Method 1 of the section on Coliform Bacteria, Enumeration of Coliforms: Determination of the Most Probable Number (p. 72).

Method 2[33]

This method includes procedures for the plate count determination, culture identification, and enrichment culturing.

I. Apparatus and materials

Note that the equipment and materials required for all of the procedures described in Method 2 are included under this one heading.

1 Requirements listed in Method 1 of the section on Preparation and Dilution of the Food Homogenate (p. 61).

2 Petri dishes, glass (100 × 15 mm) or plastic (90 × 15 mm).

3 1-ml bacteriological pipettes.

4 Water bath or air incubator for tempering agar, 45° ± 1° C.

5 Incubator, 35–37° C.

6 Colony counter (Quebec dark field model or equivalent).

7 Tally register.

8 Anaerobic jar such as the Case-Anaero jar with equipment and materials (evacuation pump and gas mixture of 10% carbon dioxide and 90% nitrogen) for obtaining anaerobic conditions.

9 Inoculating needle, preferably with nichrome or platinum-iridium wire.

10 Water baths, thermostatically controlled, one at 46° ± 0.5° C and the other at 37° ± 0.5° C.

11 Nitrate test reagents: α-Naphthylamine Solution and Sulphanilic Acid Solution. (Reagents 8 and 9, Part III).

12 Powdered zinc.

13 Reagents for staining smears for spores: (a) saturated, aqueous malachite green (approx. 7.6%); (b) 0.25% aqueous safranin.

14 Materials for staining smears by Gram's method (Reagent 1, Part III).

15 Sulphite Polymyxin Sulphadiazine Agar (SPS Agar; Medium 79, Part III), for plates.

33/This method is used widely in the United States and elsewhere and is recommended for foods in *Examination of Foods for Enteropathogenic and Indicator Bacteria: Review of Methodology and Manual of Selected Procedures* (Lewis and Angelotti, 1964).

16 Motility-Nitrate Medium (Medium 44, Part III), 10 ml volumes in 150 × 15 mm tubes.

17 Sporulation Broth (Medium 76, Part III), 10 ml volumes in 150 × 15 mm tubes.

18 Fluid Thioglycollate Medium (Medium 24, Part III), 10 ml volumes in 150 × 15 mm tubes and 25 ml volumes in 250 × 25 mm tubes.

19 Cooked Meat Enrichment Medium (Medium 16, Part III), 10 ml volumes in 150 × 15 mm tubes.

20 McClung-Toabe Egg-Yolk Agar (Medium 42, Part III), for plates.

21 Sheep-Blood Agar (Medium 73, Part III), for plates.

II. Procedure for plate count and culture identification

1 Pipette aseptically 1 ml of each dilution of the food homogenate prepared as described in Method 1 of the section on Preparation and Dilution of the Food Homogenate (p. 61) to each of appropriately marked duplicate culture dishes.

2 Pour 15–20 ml of SPS Agar into each plate, rotate and tilt to mix inoculum and agar, and allow to solidify.

3 Invert plates and place in Case-Anaero jar. Evacuate Anaero jar to 25 inches of vacuum and replace vacuum with the CO_2–N_2 gas mixture. Repeat procedure once. Place jar in a 35–37° C incubator and allow to incubate for 24 hours.

4 Following incubation, observe plates macroscopically for evidence of growth and black colony (H_2S) production.

5 Select plates showing an estimated 30–300 black colonies; and, using a colony counter with a piece of white tissue paper over the counting area, count colonies and calculate number of organisms per gram of food. This black colony count is the total clostridial count since clostridia other than *C. perfringens* may grow in this medium.

6 Select a representative number of colonies from the countable plates and inoculate a separate tube of Fluid Thioglycollate Medium with cells from each.

7 Incubate the tube cultures in a water bath at 46° ± 0.5° C for 3–4 hours.

8 Check growth in Fluid Thioglycollate Medium for purity by examining smears stained by Gram's method. Cells should appear as large Gram-positive rods with blunt ends.

9 If cultures are pure, inoculate separate tubes of Motility-Nitrate

Medium, Sporulation Broth, and Cooked Meat Medium with cells from the 3–4-hour-old fluid thioglycollate cultures.

10   Incubate the media at 37° ± 0.5° C in a water bath for 18–24 hours.

11   Examine tubes of Motility-Nitrate Medium by transmitted light for type of growth along stab. Non-motile organisms produce growth only in and along the line of stab. Motile organisms produce a diffuse growth out into the medium away from the stab.

12   Test Motility-Nitrate Medium for presence of nitrite by adding 0.5–1 ml of α-Naphthylamine Solution and the same amount of Sulphanilic Acid Solution. The production of a pink or red colour denotes the presence of nitrites. If no colour develops, mix reagents with upper third of medium by jabbing down into medium with sterile loop. (*Note*   The only known species of sulphite-reducing *Clostridium*, in addition to *C. perfringens*, which is non-motile and produces nitrite from nitrate is *C. filiforme*, an extremely rare organism described only once (Angelotti *et al.*, 1962).)

13   When desirable for confirmatory purposes, examine Sporulation Broth for spores (Bartholomew and Mittwer's "cold" method, 1950). Make a smear from sediment, air-dry, and heat-fix. Stain for 10 minutes with malachite green, wash with water, stain with aqueous safranin for 15 seconds, rinse, blot, dry, and examine microscopically. Spores will be stained green, vegetative cells red.

14   Pipette 2 ml of Sporulation Broth into a sterile test tube and heat in a 80° C water bath for 10 minutes. Remove from bath and, when cool, add 1 ml to a tube of Fluid Thioglycollate Medium. Incubate at 37° ± 0.5° C in a water bath for 18–24 hours.

15   Examine Fluid Thioglycollate Medium for evidence of growth, and observe microscopically for typical Gram-positive rods.

16   If growth is present, record that Sporulation Broth contained spores.

17   If no growth is seen, reincubate for 24 hours and examine once again. If no growth is evident after the second 24-hour incubation, record that Sporulation Broth did not contain spores.

18   The Cooked Meat stock cultures (see step 9 above) of those strains which (i) produce black colonies in SPS Agar, (ii) are non-motile and reduce nitrate, and (iii) produce spores are saved for further confirmatory tests if necessary (see IV and V of Method 1 above for procedure for serological typing and carbohydrate fermentation). In routine work, evidence obtained from the tests described in steps 4 through 18 immediately above is sufficiently reliable to enable the calculation of the plate count of *C. perfringens* to be made.

19 Calculate the total *C. perfringens* count from the percentage of the total black colony count (step 5 immediately above) that proved to be *C. perfringens* on the basis of the tests described in steps 8 through 17 above.

III. Procedure for enrichment culturing

Where it is desired to determine the presence or absence of *C. perfringens* in a food which is likely to contain a small number of cells, the following enrichment procedure is recommended.

1 Heat three 250 × 25 mm screw-capped tubes containing 25 ml of Fluid Thioglycollate Medium each in flowing steam for 10 minutes, and then cool the tubes rapidly in running tap water.

2 Into each tube place 25 g of test food.

3 Incubate one tube at 46° ± 0.5° C in a water bath for 4–6 hours and then proceed as directed in step 5 below. (*Note* Extended incubation (12–24 hours) at 46° C usually results in a reduction of the number of *C. perfringens* and overgrowth of concomitant flora capable of growing at this temperature.)

4 Place one of the remaining tubes in a boiling water bath for 1 hour and heat-shock the third by placing it in a water bath at 70° C for 10 minutes. Cool both tubes immediately after the heat treatments and incubate them at 37° ± 0.5° C in a water bath for 18–24 hours.

5 Subculture all tubes showing growth by streaking a 3 mm loopful of each on separate plates of McClung-Toabe Egg-Yolk Agar and Sheep-Blood Agar (surface-dry plates before use; see item 9, p. 158, for note on drying plates) in such a manner as to obtain separate colonies.

6 Incubate all plates anaerobically at 35–37° C for 24 hours.

7 Remove plates from anaerobic jar and hold them at room temperature for 1–4 hours.

8 Select a representative number of lecithinase-positive colonies from McClung-Toabe plates (circular, slightly raised colonies surrounded by an opaque zone) and a similar number of colonies showing complete or partial lysis. Transfer cells from each to a separate tube of Fluid Thioglycollate Medium.

9 Incubate tubes in a 46° ± 0.5° C water bath for 3–4 hours and conduct identification tests described above in II, steps 8 through 18.

Method 3[34]

This method includes procedures for the plate count determination and culture identification.

I. Apparatus and materials

1 Requirements listed for Method 2 in the section on Preparation and Dilution of the Food Homogenate (p. 63).

2 Incubator, $46° \pm 0.3°$ C.

3 Inoculating needle, preferably with nichrome or platinum-iridium wire.

4 1-ml bacteriological pipettes.

5 Petri dishes, glass (100×15 mm) or plastic (90×15 mm).

6 Thermo-regulated water bath, $60° \pm 1°$ C.

7 Anaerobic jar with equipment and materials (pyrogallol or evacuation pump and nitrogen, hydrogen, or a mixture of 10% carbon dioxide and nitrogen) for establishing anaerobic conditions in the jar.

8 Plastic pouches.[35]

9 Pouch holder.[36]

10 Materials for staining smears by Gram's method (Reagent 1, Part III).

11 Tryptone Sulphite Neomycin Agar (Medium 97, Part III), for plates.

12 Lactose Egg-Yolk Milk Agar (Medium 35, Part III), containing 250 µg of neomycin/ml, for plates.

II. Procedure

1 Pipette aseptically 1 ml of each dilution of food homogenate (prepared as described in Method 2 of the section on Preparation and Dilution of the Food Homogenate, p. 63) to a separate appropriately marked empty plastic pouch.

34/This method is used in laboratories in Continental Europe and the United States.

35/Make pouch as specified by Bladel and Greenberg (1965) or by de Waart and Smit (1967), using plastic film with an oxygen transmittancy no greater than 0.4 cc of oxygen per sq. in. per 24 hours.

36/Construct pouch holder as described by Bladel and Greenberg (1965), setting spaces for the pouches 0.5 cm apart. Pouches filled with inoculated melted agar are placed in the holder to give, on solidification, a plastic-coated agar disc 0.5 cm thick. Colonies growing anaerobically in the agar can be counted on a colony counter.

2 Add 2.3 ml of melted Tryptone Sulphite Neomycin Agar (tempered to 60° C in a water bath) to each pouch and place pouches momentarily in the pouch holder to remove entrapped air bubbles.

3 Remove pouches from holder, one at a time, and sealing the upper portion of the neck with thumb and forefinger, flex pouch three or four times with the other hand to mix contents.

4 Replace pouch in holder until agar has set, then incubate at 46° ± 0.3° C for 18–24 hours.

5 Count black colonies in all pouches containing between 20 and 200 colonies and calculate number of such colonies per gram of the original food specimen.[37]

6 For identification studies, streak cells of at least 10 typical well-isolated colonies onto the surface of separate dried plates of Lactose Egg-Yolk Milk Agar in such a manner as to obtain isolated colonies. To obtain cells, first cut away the plastic film and agar (if colony is deeply embedded) above the colony with a sterile scalpel and fish cells with a loop or needle.

7 Incubate plates anaerobically at 46° ± 0.3° C for 24 hours.

8 Prepare smears of each culture from well-isolated typical colonies and stain by Gram's method.

9 Use cells of cultures composed of short, thick Gram-positive rods occurring singly, in pairs, or in short chains to test for the Nagler reaction by procedure described in Method 1 above (III, step 5).

10 From the percentage of isolates giving a typical colonial appearance on Lactose Egg-Yolk Milk Agar and a positive Nagler reaction, calculate the number of cells of *C. perfringens* in the original food specimen from the anaerobic plate count obtained in II, step 5 above.

11 If identification of the serotypes involved is desired, follow procedures outlined in Method 1 above.

Bacillus cereus[38]

Bacillus cereus is enumerated in foods in order to investigate its involvement, or potential involvement, in food-poisoning. The organism has little other specific connotation except to indicate that if it is present in the numbers necessary to cause food-poisoning (10^7 or greater per gram), the food could not have been continuously refrigerated.

37/*C. bifermentans*, which will give black colonies, is prevented from growing at 46° C by the neomycin (50 μg/ml) present in the medium.
38/This method follows closely that described by Mossel *et al.* (1966, 1967).

In the following method *B. cereus* is enumerated on a selective medium (Phenol-Red Egg-Yolk Polymyxin Agar) containing mannitol. Presumptive *B. cereus* colonies are surrounded by a halo of precipitate, due to lecithinase formation, and show a distinctly red background, due to inability of the organism to dissimilate mannitol.

ENUMERATION AND CULTURE IDENTIFICATION

I. Apparatus and materials

1 Requirements listed for Method 2 in the section on Preparation and Dilution of the Food Homogenate (p. 63).

2 Petri dishes, glass (100 × 15 mm) or plastic (90 × 15 mm).

3 1-ml bacteriological pipettes with subdivisions of 0.1 ml or less.

4 Water bath or air incubator for tempering agar, 45° ± 1° C.

5 Incubator, 30° ± 1° C.

6 Materials for staining smears by Gram's method (Reagent 1, Part III).

7 Phenol-Red Egg-Yolk Polymyxin Agar (Medium 61, Part III), for plates.

8 Phenol-Red Carbohydrate Broths for the following carbohydrates: glucose, sucrose, glucerol, and salicin (see Medium 60, Part III), 5 ml volumes in tubes.

9 Nitrate Broth (Medium 45, Part III), 5 ml volumes in tubes.

10 Starch Agar (Medium 78, Part III), for plates.

11 Litmus Milk (Medium 37, Part III), 10 ml volumes in tubes.

12 Buffered Glucose Broth (Medium 14, Part III), 5 ml volumes in tubes.

13 Nutrient Gelatin (Medium 48, Part III), 4–5 cm depth in tubes.

14 Nutrient Agar (Medium 46, Part III), slopes.

15 Lugol's Iodine Solution (Reagent 5, Part III).

16 Sulphanilic Acid Solution (Reagent 9, Part III).

17 α-Naphthylamine Solution (Reagent 8, Part III).

18 Creatine crystals.

19 40% aqueous solution of potassium hydroxide.

20 Naphthol Solution (Reagent 7, Part III).

II. Procedure

1 Prepare food samples by the procedure described in Method 2 of the section on Preparation and Dilution of the Food Homogenate (p. 63).

Prepare dilutions of 10^{-1}, 10^{-2}, 10^{-3}, and 10^{-4}.

2 Streak evenly 0.1 ml of each of the dilutions over the dried surface of a separate plate of Phenol-Red Egg-Yolk Polymyxin Agar (see item 9, p. 158, for note on drying agar plates).

3 Incubate aerobically at 30° C for 45 hours.

4 Count the colonies surrounded by a halo of dense precipitate (lecithinase activity) with a distinct violet-red background and calculate the total count per gram of food specimen. This is the presumptive *B. cereus* count.

5 If it is necessary to confirm the colonies as *B. cereus*, select at random a number of typical colonies equal to the square root of the number obtained on the counted plates.

6 Purify these by streaking each on a separate plate of Nutrient Agar in such a manner as to obtain isolated colonies. Incubate plates at 30° C for 24 hours.

7 Transfer cells from an isolated colony on each plate to a separate slope of Nutrient Agar. Incubate the tubes at 30° C for 24 hours.

8 Prepare smears from slope cultures and stain by Gram's method. *B. cereus* will appear as short Gram-positive rods with square ends, in short to long tangled chains. The spores are ellipsoidal, central to subcentral, thin-walled, and do not swell the cell.

9 Inoculate the following determinative media with 24-hour cells from Nutrient Agar slopes in the manner described:

(*a*) *Phenol-Red Carbohydrate Broth*

(i) Inoculate a loopful of cells into tubes of each of the following carbohydrates: glucose, sucrose, glycerol, and salicin. Incubate tubes at 30° C for 24 hours.

(ii) Record acid and gas production. *B. cereus* ferments all of the carbohydrates with production of acid (medium turns yellow) but no gas.

(*b*) *Nitrate Broth*

(i) Inoculate with a loopful of cells and incubate tubes at 30° C for 18–24 hours.

(ii) Add ½–1 ml each of Sulphanilic Acid Solution and α-Naphthylamine Solution to each culture. Shake the tubes and observe the colour. Development of a distinct pink or red colour represents a positive nitrate reduction test. *B. cereus* reduces nitrate.

(*c*) *Litmus Milk*

(i) Inoculate with a loopful of cells and incubate tubes at 30° C for 24 hours.

(ii) Record reaction. *B. cereus* causes rapid peptonization with or without slight coagulation.

(*d*) *Nutrient Gelatin*

(i) Stab inoculate with a needle and incubate tubes at 30° C for 24 hours in an upright position.

(ii) Record reaction. *B. cereus* rapidly liquefies gelatin.

(*e*) *Starch Agar*

(i) Pour plates of Starch Agar, dry surfaces, and inoculate by making one streak with a loopful of cells across the full width of the plate. Two or three cultures can be tested on the same plate. Incubate at 30° C for 24 hours.

(ii) Flood the surface of plates with Lugol's Iodine Solution. A clear zone around the streaks indicates starch hydrolysis. *B. cereus* hydrolyses starch.

(*f*) *Buffered Glucose Broth*

(i) Inoculate with a loopful of cells and incubate at 30° C for 48 hours.

(ii) Test for production of acetylmethylcarbinol by pipetting 1 ml of culture to an empty tube and adding 0.6 ml of a naphthol solution and 0.2 ml of 40% potassium hydroxide. Shake tubes after addition of each reagent. A few crystals of creatine added to the test medium will intensify and speed reaction. Perform test at room temperature and read results 4 hours after addition of reagents. Development of a pink colour is a positive test. *B. cereus* gives a positive test.

10 From the percentage of isolates proving to be *B. cereus* on the basis of the above tests, calculate the number of cells in the original food specimen from the total colony count in step 4 above.

REFERENCES

AMERICAN PUBLIC HEALTH ASSOCIATION. 1960. Standard methods for the examination of dairy products (11th ed.; New York: American Public Health Association, Inc.).

AMERICAN PUBLIC HEALTH ASSOCIATION, AMERICAN WATER WORKS ASSOCIATION, and WATER POLLUTION CONTROL FEDERATION. 1965. Standard methods for the examination of water and wastewater (12th ed.; New York: American Public Health Association, Inc.).

AMERICAN PUBLIC HEALTH ASSOCIATION, SUBCOMMITTEE ON METHODS FOR THE MICROBIOLOGICAL EXAMINATION OF FOODS. 1966. Recommended methods for the microbiological examination of foods (2nd ed.; New York: American Public Health Association, Inc.).

ANGELOTTI, R. 1964. Significance of "total counts" in bacteriological examination of foods. *In* Lewis and Angelotti (1964), p. 5.

ANGELOTTI, R., HALL, H. E., FOTER, M. J., and LEWIS, K. H., 1962. Quantitation of *Clostridium perfringens* in foods. Appl. Microbiol. *10*, 193.

ASSOCIATION OF FOOD AND DRUG OFFICIALS OF THE UNITED STATES, ADVISORY COMMITTEE ON THE MICROBIOLOGY OF FROZEN FOODS. 1966. Microbiological examination of precooked frozen foods. Bull. Assoc. Food Drug Officials U.S., Suppl. Issue.

ASSOCIATION OF OFFICIAL ANALYTICAL CHEMISTS. 1967. Detection and identification of *Salmonella* in egg products. J. Assoc. Offic. Anal. Chemists *50*, 231.

BAER, E. F. 1966. Proposed method for isolating coagulase-positive staphylococci from food products: Report of a collaborative study. J. Assoc. Offic. Anal. Chemists *49*, 270.

BAER, E. F., FRANKLIN, M. K., and GILDEN, M. M. 1966. Efficiency of several selective media for isolating coagulase-positive staphylococci from food products. J. Assoc. Offic. Anal. Chemists *49*, 267.

BAIRD-PARKER, A. C. 1962a. An improved diagnostic and selective medium for isolating coagulase-positive staphylococci. J. Appl. Bacteriol. *25*, 12.

—— 1962b. The performance of an egg yolk-tellurite medium in practical use. J. Appl. Bacteriol. *25*, 441.

BARNES, E. M. 1956. Methods for the isolation of faecal streptococci (Lancefield Group D) from bacon factories. J. Appl. Bacteriol. *19*, 193.

BARRAUD, C., KITCHELL, A. G., LABOTS, H., REUTER, G., and SIMONSEN, B. 1967. Standardization of the total aerobic count in meat and products. Fleischwirtschaft *47*, 1313.

BARTHOLOMEW, J. W. and MITTWER, T. 1950. A simplified bacterial spore stain. Stain Technol. *25*, 153.

BLADEL, B. O. and GREENBERG, R. A. 1965. Pouch method for the isolation and enumeration of clostridia. Appl. Microbiol. *13*, 281.

BREED, R. S., MURRAY, E. G. D., and NATHAN, R. S. 1957. Bergey's Manual of determinative bacteriology (7th ed.; Baltimore, Md.: The Williams and Wilkins Co.).

BUCK, J. D. and CLEVERDON, R. C. 1960. The spread plate as a method for the enumeration of marine bacteria. Limnology and Oceanography *5*, 78.

BUTTERFIELD, C. T. 1932. The selection of dilution waters for bacteriological examinations. J. Bacteriol. *23*, 355.

BUTTIAUX, R., SAMAILLE, J., and PIERENS, Y. 1956. L'identification des Escherichaea des eaux. Test d'Eykman et production d'indole a 44° C. Tests IMViC. Ann. Inst. Pasteur, Lille *8*, 137.

CAMPBELL, J. J. R. and KONOWALCHUK, J. 1948. Comparison of "drop" and "pour" plate counts of bacteria in raw milk. Can. J. Res. *26E*, 327.

CLARK, D. S. 1967. Comparison of pour and surface plate methods for determination of bacterial counts. Can. J. Microbiol. *13*, 1409.

COSTIN, I., DAVID, P., DINCULESCU, M., WEISZBERGER, A., OLARIU, GH., SCHIFTEN, N., and MARTON, T. 1960. Toxiinfectie alimentara cu germeni din genul *Escherichia*. Microbiol. Parazitol. Epidemiol. (Bucharest) *6*, 531.

COSTIN, I. D., VOICULESCU, D., and GORCEA, V. 1964. An outbreak of food poisoning in adults associated with *Escherichia coli* serotype 86: B_7H_{84}. Pathol. Microbiol. *27*, 68.

CRISLEY, F. D., ANGELOTTI, R., and FOTER, M. J. 1964. Multiplication of *Staphylococcus aureus* in synthetic cream fillings and pies. Public Health Rept. (U.S.) *79*, 369.

CRISLEY, F. D., PEELER, J. T., and ANGELOTTI, R. 1965. Comparative evaluation of five selective and differential media for the detection and enumeration of coagulase-positive staphylococci in foods. Appl. Microbiol. *13*, 140.

DE MELLO, C. G., DANIELSON, I. S., and KISER, J. S. 1951. The toxic effect of buffered saline solutions on the viability of *Brucella abortus*. J. Lab. Clin. Med. *37*, 577.

DEPARTMENT OF HEALTH AND WELFARE, JAPAN. 1963. Methods for the examination of *V. parahaemolyticus*. In *Vibrio parahaemolyticus*, edited by T. Fujino and H. Fukumi (Tokyo: Nayashoten), pp. 400–22 (Text in Japanese).

DE WAART, J. and SMIT, F. 1967. The enumeration of obligately anaerobic bacteria using pouches made from plastics with a low oxygen permeability. Lab. Pract. *16*, 1098.

DOLMAN, C. E. 1957. Recent observations on type E botulism. Can. J. Public Health *48*, 187.

——— 1964. Growth and metabolic activities of *C. botulinum* types. In Botulism, Proc. of Symp., Cincinnati, Ohio, Jan. 13–15, 1964, edited by K. H. Lewis and K. Cassel, Jr. (U.S. Public Health Service, Publ. No. 999-FP-1 [Cincinnati, Ohio: Public Health Service]), p. 43.

DUNCAN, A. J. 1965. Quality control and industrial statistics (3rd ed.; Homewood, Ill.: Irwin).

144 MICROORGANISMS IN FOODS

EDWARDS, P. R. and EWING, W. H. 1962. Identification of *Enterobacteriaceae* (2nd ed.; Minneapolis: Burgess Pub. Co.).

EIJKMAN, C. 1904. Die Gärungsprobe bei 46° C als Hilfsmittel bei der Trinkwasseruntersuchung. Zentr. Bakteriol. Parasitenk., I, Orig. *37*, 742.

ENTEROBACTERIACEAE SUBCOMMITTEE OF THE NOMENCLATURE COMMITTEE OF THE INTERNATIONAL ASSOCIATION OF MICROBIOLOGICAL SOCIETIES. 1963. Report of 1962 Meeting, Montreal. Int. Bull. Bacteriol. Nomencl. Taxon. *13*, 141.

EWING, W. H., DAVIS, B. R., and MONTAGUE, T. S. 1963. Studies on the occurrence of *Escherichia coli* serotypes associated with diarrhoeal diseases (Atlanta, Ga.: Communicable Disease Center, Laboratory Branch, U.S. Department of Health, Education and Welfare).

EWING, W. H., TATUM, N. W., and DAVIS, B. R. 1957. The occurrence of *Escherichia coli* serotypes associated with diarrhoeal disease in the United States. Public Health Lab. *15*, 118.

FISHBEIN, M. 1961. The aerogenic response of *Escherichia coli* and strains of *Aerobacter* in EC broth and selected sugar broths at elevated temperatures. Appl. Microbiol. *10*, 79.

FISHBEIN, M. and SURKIEWICZ, B. F. 1964. Comparison of the recovery of *Escherichia coli* from frozen foods and nutmeats by confirmatory incubation in EC medium at 44.5 and 45.5° C. Appl. Microbiol. *12*, 127.

FISHBEIN, M., SURKIEWCZ, B. F., BROWN, E. F., OXLEY, H. M., PADRON, A. P., and GROOMES, R. J. 1967. Coliform behavior in frozen foods. I. Rapid test for recovery of *Escherichia coli* from frozen foods. Appl. Microbiol. *15*, 233.

GAUDY, A. F., JR., ABU-NIAAJ, F., and GAUDY, E. T. 1963. Statistical study of the spot-plate technique for viable-cell counts. Appl. Microbiol. *11*, 305.

GELDREICH, E. E. 1966. Sanitary significance of fecal coliforms in the environment. (U.S. Department of the Interior, Federal Water Pollution Control Administration, Washington, D.C., Publ. WP-20-3).

GIBBS, B. M. and FREAME, B. 1965. Methods for the recovery of clostridia from foods. J. Appl. Bacteriol. *28*, 95.

GUINÉE, P. A. M. and MOSSEL, D. A. A. 1963. The reliability of the test of McKenzie, Taylor and Gilbert for detection of faecal *Escherichia coli* strains of animal origin in foods. Antonie van Leeuwenhoek, J. Microbiol. Serol. *29*, 163.

GUNTER, S. E. 1954. Factors determining the viability of selected microorganisms in inorganic media. J. Bacteriol. *67*, 628.

HALL, H. E. 1964. Methods of isolation and enumeration of coliform organisms. *In* Lewis and Angelotti (1964).

HARTMAN, P. A. and HUNTSBERGER, D. V. 1961. Influence of subtle differences in plating procedure on bacterial counts of prepared frozen foods. Appl. Microbiol. *9*, 32.

HAUSCHILD, A. H. W., ERDMAN, I. E., HILSHEIMER, R., and THATCHER, F. S. 1967. Variations in recovery of *Clostridium perfringens* on commercial sulfite-polymyxin-sulfadiazine (SPS) agar. J. Food Sci. *32*, 469.

HOBBS, B. C. 1965. *Clostridium welchii* as a food poisoning organism. J. Appl. Bacteriol. *28*, 74.

HOBBS, B. C., SMITH, M. E., OAKLEY, C. L., WARRACK, G. H., and CRUICKSHANK, J. C. 1953. *Clostridium welchii* food poisoning. J. Hyg., Camb. *51*, 75.

INTERNATIONAL DAIRY FEDERATION. 1958. International standard FIL-IDF 3:1958. Colony count of liquid milk and dried milk (General Secretariat 10, rue Ortelius, Brussels 4).

INTERNATIONAL ENTEROBACTERIACEAE SUBCOMMITTEE REPORT. 1958. Int. Bull. Bacteriol. Nomencl. Taxon. 8, 25, 93.

JOHNSON, H. M., BRENNER, K., ANGELOTTI, R., and HALL, H. E. 1966. Serological studies of types A, B and E botulinal toxins by passive haemagglutination and bentonite flocculation. J. Bacteriol. 91, 967.

KING, W. L. and HURST, A. 1963. A note on the survival of some bacteria in different diluents. J. Appl. Bacteriol. 26, 504.

KODAMA, T. 1967. On the haemolytic activity of V. parahaemolyticus. In Vibrio parahaemolyticus, Series II, edited by T. Fujino and H. Fukumi (Tokyo: Nayashoten), p. 232 (Text in Japanese).

KORETSKAIA, L. S. and KOVALEVSKAIA, A. N. 1958. Food poisonings produced by B. coli serotype $O_{26}:B_8$. Zh. Mikrobiol, Epidemiol. i Immunobiol. 4, 58 (Eng. transl. in J. Microbiol., Epidemiol., Immunobiol. (USSR) 29, 553).

KOSER, S. A. 1923. Utilization of the salts of organic acids by the colon-aerogenes group. J. Bacteriol. 8, 493.

KOVACS, N. 1928. A simplified method for detecting indol formation by bacteria. Z. Immunitätsforsch. 56, 311; Chem. Abs. 22, 3425.

―――― 1956. Identification of Pseudomonas pyocyanae by the oxidase reaction. Nature 178, 703.

LEVINE, M. 1916. On the significance of the Voges-Proskauer reaction. J. Bacteriol. 1, 153.

LEWIS, K. H. and ANGELOTTI, R. (eds.). 1964. Examination of foods for enteropathogenic and indicator bacteria. Review of methodology and manual of selected procedures (Division of Environmental Engineering and Food Protection, U.S. Public Health Service, Publ. No. 1142).

LJUTOV, V. 1961. Technique of methyl red test. Acta Pathol. Microbiol. Scand. 51, 369.

―――― 1963. Technique of Voges-Proskauer test. Acta Pathol. Microbiol. Scand. 58, 325.

LUNDBECK, H. and TIRUNARAYANAN, M. O. 1966. Investigations on the enzymes and toxins of staphylococci. Study of the "egg yolk reaction" using an agar plate assay method. Acta Pathol. Microbiol. Scand. 68, 123.

MACKENZIE, E. F. W., TAYLOR, E. W., and GILBERT, W. E. 1948. Recent experiences in the rapid identification of Bacterium coli Type I. J. Gen. Microbiol. 2, 197.

MCCLUNG, L. S. and TOABE, R. 1947. The egg yolk plate reaction for the presumptive diagnosis of Clostridium sporogenes and certain species of the gangrene and botulinum groups. J. Bacteriol. 53, 139.

MILES, A. A. and MISRA, S. S. 1938. The estimation of the bactericidal power of the blood. J. Hyg., Camb. 38, 732.

MINISTRY OF HEALTH AND MINISTRY OF HOUSING AND LOCAL GOVERNMENT. 1956. The bacteriological examination of water supplies (3rd ed.; Public Health and Medical Subjects Report No. 71 [London: H.M.S.O.]).

MOSSEL, D. A. A., DE BRUIN, A. S., VAN DIEPEN, H. M. J., VENDRIG, C. M. A., and ZOUTEWELLE, G. 1956. The enumeration of anaerobic bacteria, and of

Clostridium species in particular, in foods. J. Appl. Bacteriol. *19*, 142.

MOSSEL, D. A. A., KOOPMAN, M. J., and JONGERIUS, E. 1966. The enumeration of *Bacillus cereus* in foods. Antonie van Leeuwenhoek, J. Microbiol. Serol. *32*, 453.

———— 1967. Enumeration of *Bacillus cereus* in foods. Appl. Microbiol. *15*, 650.

MOSSEL, D. A. A., MENGERINK, W. H. J., and SCHOLTS, H. H. 1962. Use of a modified MacConkey agar medium for the selective growth and enumeration of *Enterobacteriaceae*. J. Bacteriol. *84*, 381.

MOSSEL, D. A. A., VISSER, M., and CORNELISSEN, A. M. R. 1963. The examination of foods for *Enterobacteriaceae* using a test of the type generally adopted for the detection of salmonellae. J. Appl. Bacteriol. *26*, 444.

NARAYAN, K. G. 1967. Culture isolation and identification of clostridia. Zentr. Bakteriol., I, Orig. *202*, 212.

NEFEDJEVA, M. P. 1964. Laboratornaia diannostika infektsionnykh zabolevanii [Laboratory diagnosis of infectious diseases; methodological manual] (2nd ed.; Moscow: Biuro Nauchnoi Informatsii), p. 352.

NETER, E., WEBB, C. R., SHUMWAY, C. N., and MURDOCK, M. R. 1951. Study on etiology, epidemiology and antibiotic therapy of infantile diarrhoea, with a particular reference to certain serotypes of *Escherichia coli*. Am. J. Public Health *41*, 1490.

NORTH, W. R., JR. 1960. Use of crystal violet or brilliant green dyes for the determination of salmonellae in dried food products. J. Bacteriol. *80*, 861.

PACKER, R. A. 1943. The use of sodium azide (NaN_3) and crystal violet in a selective medium for streptococci and *Erysipelothrix rhusiopathiae*. J. Bacteriol. *46*, 343.

REED, R. W. and REED, G. B. 1948. "Drop plate" method of counting viable bacteria. Can. J. Res. *26E*, 317.

SAKAZAKI, R. 1967. Isolation and identification of *Vibrio parahaemolyticus*. *In Vibrio parahaemolyticus*, Series II, edited by T. Fujino and H. Fukumi Tokyo: Nayashoten), p. 119 (Text in Japanese).

SHAW, D. B. and WILSON, J. B. 1963. Egg yolk factor of *Staphylococcus aureus*. I. Nature of the substrate and enzyme involved in the egg yolk opacity reaction. J. Bacteriol. *85*, 516.

SHELTON, L. R., LEININGER, H. V., SURKIEWICZ, B. F., BAER, E. F., ELLIOTT, R. P., HYNDMAN, J. B., and KRAMER, N. 1962. A bacteriological survey of the frozen precooked food industry (Washington, D.C.: U.S. Department of Health, Education and Welfare, Food and Drug Administration).

SHERMAN, J. M. 1937. The streptococci. Bacteriol. Rev. *1*, 3.

SHEWAN, J. M. and BAINES, C. R. 1963. A comment on sampling techniques for microbiological detection of salmonellae in consignments of foods and feeds. Sampling of fish and fishery products for bacterial infection. *In* Radiation control of salmonellae in food and feed products. (IAEA Tech. Rept. Series, No. 22), p. 133.

SIMMONS, J. S. 1926. A culture medium for differentiating organisms of the typhoid-colon-aerogenes groups and for the isolation of certain fungi. J. Infect. Diseases *39*, 209.

SOCIETY FOR APPLIED BACTERIOLOGY. 1966. Identification methods for micro-biologists, edited by B. M. Gibbs and F. A. Skinner (Technical Series No. 1, Part A [London: Academic Press]).

STOKES, J. L. and OSBORNE, W. W. 1956. Effect of the egg shell membrane on bacteria. Food Res. *21*, 264.

STRAKA, R. P. and STOKES, J. L. 1957. Rapid destruction of bacteria in commonly used diluents and its elimination. Appl. Microbiol. *5*, 21.

SUBCOMMITTEE ON SAMPLING AND METHODOLOGY FOR SALMONELLAE IN EGGS, EGG-PRODUCTS AND CAKE MIXES. 1966. Recommendations for sampling and laboratory analysis of eggs, egg-products, and prepared mixes. Food Technol. *20*, 121.

TAYLOR, J. and CHARTER, R. E. 1952. The isolation of serological types of *Bact. coli* in two residential nurseries and their relation to infantile gastro-enteritis. J. Pathol. Bacteriol. *64*, 715.

TAYLOR, W. I. 1965. Isolation of shigellae. I. Xylose lysine agars; new media for isolation of enteric pathogens. Am. J. Clin. Pathol. *44*, 471.

THATCHER, F. S. 1963. The microbiology of specific frozen foods in relation to public health: Report of an international committee. J. Appl. Bacteriol. *26*, 266.

THOMAS, G. and CHEFTEL, H. 1955. Prélèvement d'échantillons en vue de l'appréciation de la qualité bactériologique de lots de conserve. Annales des falsifications et des fraudes *48*, 223, 283.

TIRUNARAYANAN, M. O. and LUNDBECK, H. 1967. Investigations on the enzymes and toxins of staphylococci. Acta Pathol. Microbiol. Scand. *69*, 314.

TURNER, F. J. and SCHWARTZ, B. S. 1958. The use of a lyophilized human plasma standardized for blood clotting factors in the coagulase and fibrino-lytic tests. J. Lab. Clin. Med. *52*, 888.

UNITED STATES FOOD AND DRUG ADMINISTRATION. 1966. Bacteriological Analytical manual (Washington, D.C.: Food and Drug Administration, Department of Health, Education and Welfare).

VAN SCHOTHORST, M., MOSSEL, D. A. A., KAMPELMACHER, E. H., and DRION, E. G. 1966. The estimation of the hygienic quality of feed components using an Enterobacteriaceae enrichment test. Zentr. Veterinaermed. *13B*, 273.

VOGEL, R. A. and JOHNSON, M. 1960. A modification of the tellurite-glycine medium for use in the identification of *Staphylococcus aureus*. Public Health Lab. *18*, 131.

WINSLOW, C. E. A. and BROOKE, O. R. 1927. The viability of various species of bacteria in aqueous suspensions. J. Bacteriol. *13*, 235.

PART III

SPECIFICATIONS FOR
MEDIA, REAGENTS,
AND INGREDIENTS

SPECIFICATIONS FOR INGREDIENTS FOR MEDIA AND REAGENTS

For a given method of examination of foods to have similar significance wherever it is used, it is essential that the media, ingredients, and testing reagents be of comparable standard. The following compilation of specifications for media and test reagents was thus undertaken to permit any laboratory unable to obtain a specific commercial medium to use an alternative medium or to prepare one of its own to appropriate specifications. It is hoped that it will be of value, particularly to those countries where certain commercial media may not be available because of the absence of an outlet for distribution or because of restrictions on currency movements or a shortage of budgetary funds for their purchase.

Many of the ingredients listed are trade names used by companies to indicate a particular product. The source and order number are given in parentheses only where it is understood that no other manufacturer makes a product by the same name. It is believed, however, that in most instances a similar product is made by other companies under another name and that several of the commercial names listed here refer to similar products. No attempt is made to list comparable or alternative preparations; only those products which were specifically prescribed for the preparation of one or more of the media listed in the section on Media Formulae and Directions for Preparation are included.

1 AGAR

Only bacteriological grade agar should be used.

2 BEEF EXTRACT

Any standardized brand of beef extract especially manufactured for the preparation of microbiological culture media may be used. Meat infusion is not satisfactory.

3 BILE SALTS

Bile salts is a standardized mixture of sodium glycocholate and sodium taurocholate, prepared from fresh ox-bile, for use as a selectively inhibitory agent in media. Use only standardized commercially available dehydrated preparations especially prepared for bacteriological work.

4 BILE SALTS NO. 3

A refined bile salt, it is used as a selectively inhibitory agent in bacteriological culture media in amounts less than one third of the concentration of bile salts normally quoted in formulae for bile salt media. Use only commercially available dehydrated preparations especially prepared for bacteriological work.

5 SUGARS

All sugars used in the preparation of culture media must be chemically pure and known to be suitable for bacteriological purposes.

6 CASITONE

Casitone is a pancreatic digest of casein and is a rich source of amino acid nitrogen. It has many uses in cultivation media. To provide uniformity of culture media, dehydrated, commercially available casitone (Difco B259) should be used.

7 GENERAL CHEMICALS (SALTS)

All general chemicals used as ingredients in culture media must be ACS (American Chemical Society) or AR (analytical reagent) grade or equivalent.

8 DYES

Only dyes certified by the Commission on Standardization of Biological Stains for use in the preparation of culture media should be employed.

9 GELATIN

Gelatin is a refined water-soluble proteaceous material free from fermentable carbohydrates, used for solidification of culture media and for the detection and differentiation of certain proteolytic bacteria. Use only products which have been especially prepared for bacteriological use.

10 GELYSATE

Gelysate is a pancreatic hydrolysate of gelatin characterized by a low cystine, tryptophane, and carbohydrate content. The commercially available dehydrated preparation (Baltimore Biol. Lab. 02–190) is recommended for uniformity of culture media.

11 NEOPEPTONE

Neopeptone is an enzymic protein digest suitable for use in the propagation of organisms considered difficult to cultivate *in vitro*. To provide uniformity of culture media, dehydrated commercially available neopentone (Difco B119) is recommended.

12 OX-GALL

Ox-gall is fresh bile purified for use in culture media as a selectively inhibitory agent in a number of bile media. Only standardized commercially available dehydrated ox-gall prepared for bacteriological use is recommended. An 8–10% aqueous solution of the dehydrated product is equivalent to fresh bile.

13 PEPTONE

This is a general purpose peptone recommended for the preparation of routine bacteriological media. Any peptone which comparative tests have shown to give satisfactory results may be used. To provide uniformity of culture media, dehydrated commercially available peptone is recommended.

14 PHYTONE

Phytone is a vegetable peptone prepared by the papain digestion of soya bean meal. It is used in media for the cultivation of fastidious organisms where rapid and profuse growth is required. Dehydrated commercially available phytone (Baltimore Biol. Lab. 02–144) is recommended, to provide uniformity of culture media.

15 POLYPEPTONE

Polypeptone is a mixture of peptones, made up of equal parts of Trypticase (Ingredient 19) and Thiotone (Ingredient 18) for use in media where the characteristics of both peptones are desirable. To provide uniformity of culture media, dehydrated commercially available polypeptone (Baltimore Biol. Lab. 02–149) is recommended.

16 PROTEOSE PEPTONE

This is a specialized peptone prepared by the papain digestion of selected fresh meat for use in media mainly for the production of bacterial toxins. Any proteose peptone preparation which comparative tests have shown to give satisfactory results can be used. It is recommended that a dehydrated commercially available source be used to provide uniformity of culture media.

17 SKIM MILK POWDER

Any standardized brand of thermophile-free powdered skim milk especially manufactured for the preparation of microbiological culture media may be used. Ordinary skim milk powder is not satisfactory.

18 THIOTONE

Thiotone is prepared by pectic digestion of animal tissues and is characterized by a high sulphur content making it useful in testing for hydrogen sulphide formation. To provide uniformity of culture media, the commercially available dehydrated product (Baltimore Biol. Lab. 02–108) is recommended.

19 TRYPTICASE

This is a peptone derived from casein by pancreatic digestion. A rich source of amino acid nitrogen, it has many uses in cultivation media. To provide uniformity of culture media, dehydrated commercially available (Baltimore Biol. Lab. 02-148) trypticase is recommended.

20 TRYPTONE

Tryptone is a pancreatic or trypsinic hydrolysate of high-grade casein and a rich source of amino acid nitrogen. Any tryptone which comparative tests have shown to give satisfactory results can be used. To provide uniformity of culture media, dehydrated commercially available tryptone is recommended.

21 TRYPTOSE

Tryptose is a mixed peptone with nutritional properties making it suitable for use in media for the isolation and cultivation of fastidious organisms. Any tryptose preparation which comparative tests have shown to give satisfactory results may be used. To provide uniformity of culture media, dehydrated commercially available tryptose is recommended.

22 YEAST EXTRACT

Yeast extract is the water-soluble extract of lysed yeast cells. It is an excellent source of growth-stimulatory substances. Any yeast extract shown to yield satisfactory results may be used. To provide uniformity of culture media, dehydrated commercially available yeast extract is recommended.

23 WATER

Only distilled or demineralized water which has been tested and found non-toxic to microorganisms should be used for the preparation of culture media. The biological test procedure prescribed by the American Public Health Association (1960) and the American Public Health Association *et al.* (1965) is recommended. It is based on the growth of *Enterobacter aerogenes* in a chemically defined minimal growth medium. The effect of the addition of toxic agents, such as those which might be contained in distilled water, is measured in terms of the population density.

SPECIFICATIONS FOR
MEDIA

This chapter contains the formulae for all culture media mentioned in Part II of this book and directions for preparing the media. In the text of the methods in Part II the recommended medium is referred to by name followed by the medium number of this chapter in parentheses: e.g., Endo Agar (Medium 21, Part III).

Unless otherwise stated the use of commercially available dehydrated media is recommended for convenience and to provide uniformity of preparation. If media are prepared from basic ingredients, the directions outlined herein should be followed.

Similar basic procedural steps are followed in the preparation of most of the media:

1   The ingredients are added in the correct amounts to distilled water (at room temperature) in a suitable container, preferably of pyrex glass or stainless steel. Ingredients required in small concentration or having low solubilities are often more conveniently added as filtered aqueous solutions or, if the solubility in water is low, as alcoholic or alkali solutions. In some special instances certain ingredients (e.g., some carbohydrates, egg-yolk suspension, and sodium sulphite) must be prepared apart from the main bulk of the medium, often sterilized separately by filtration (see step 8 below) and then added, with aseptic precaution, to the portion sterilized in the autoclave. This procedure is necessary to avoid undesirable degradations or reactions which would otherwise take place during normal autoclaving.

2   To assist with the solution of the ingredients, particularly if one or more of them is a dehydrated product, it is good practice to permit the mixture to soak about 15 minutes. This reduces the amount of heating necessary to obtain complete solution (step 3 below) and hence prevents unnecessary evaporation and decomposition.

3   The ingredients are then dissolved completely with minimal delay either (*a*) by heating, to boiling if agar is included in the medium, with

frequent agitation, above an asbestos-centred wire gauze over a burner, or (b) by exposing the suspension to flowing steam in a steamer.

4 The medium is next cooled to approximately room temperature, if it is a broth medium, or to about 50° C if it is an agar medium. Readjustment of the volume with distilled water (warmed to 45–50° C for agar media) can be carried out at this point if a significant loss has occurred during heating. However, unless the heating period has been extended unduly, evaporation loss will be negligible (less than 1%) and no adjustment is necessary.

5 If the medium contains inherent precipitates or undissolved materials and if these substances interfere with the intended use of the medium, it must be clarified. Any method of clarification which yields a medium suitable for detection of all bacterial colonies and which will not remove or add nutritive ingredients can be used. Clarification can be accomplished by centrifugation, by sedimentation, or, in the case of melted agar, by filtration through cotton over cheese-cloth or through towels or coarse filter paper.

6 The reaction of the clarified medium must be adjusted to a predetermined value so that the desired final pH will be obtained after autoclaving. For most unbuffered media autoclaving will lower the pH by 0.1–0.2 unit but occasionally the drop will be as great as 0.4. When buffering salts, such as phosphates, are present in the media, the decrease in pH value as determined will be negligible. Measurement of pH should preferably be made with a pH meter. If a meter is not available, follow directions for the colorimetric determination of pH given in *Standard Methods for the Examination of Water and Wastewater* (APHA et al., 1965).

7 The medium is then distributed into tubes, flasks, or bottles as required. If plates are to be poured immediately after autoclaving (see step 9 below) the medium can be sterilized in the flask in which it was prepared.

8 With a number of exceptions media are sterilized by autoclaving at 121° C for 15 minutes or at 116° C for 20 minutes. Carbohydrate broths, if sterilized by heat, should be autoclaved at 121° C for 5 minutes or 116° C for 10–12 minutes. As mentioned (step 1 above), it is sometimes preferable to sterilize solutions of carbohydrates or substances with other specifications separately by filtration through an asbestos pad, a membrane filter, or a filter candle of suitable porosity. The sterile solution is then added aseptically to the base portion of the medium previously sterilized by autoclaving.

9   In many cases freshly prepared and sterilized medium is poured into Petri dishes for immediate use in surface culturing. The plates should, however, be dried before they are inoculated to prevent the spreading and confluence of the colonies. This can be accomplished by placing the plates (*a*) in a convection-type oven or incubator at 50° C for 30 minutes with lids removed and agar surface downwards, (*b*) in an oven or incubator (preferably a forced-air type) for 2 hours at 50° C with lids on and agar surface upwards, (*c*) in a 37° C incubator for 4 hours with lids on and agar surface upwards, or (*d*) on the laboratory bench for about 16 hours at room temperature with lids on and agar surface upwards.

Media formulae and directions for preparation[1]

1 BAIRD-PARKER AGAR (Baird-Parker, 1962*a, b*)

Formula for basal medium

Tryptone[2]	10	g
Beef extract[3]	5	g
Yeast extract[4]	1	g
Lithium chloride, hexahydrate	5	g
Agar	20	g
Sodium sulphamezathine	0.055	g

Directions for preparation of basal medium   Add above ingredients to 1 litre of distilled water and heat with agitation to obtain complete solution, cool to 50–60° C, and adjust pH to 6.8. Dispense unfiltered in 90 ml amounts in bottles, and sterilize at 121° C for 15 minutes. The pH, after autoclaving, should be 6.8–7.0.

Note that if the food specimen contains a large number of *Proteus*, 0.055 g of sulphamezathine can be added as an inhibitor to each litre of the basal medium prior to sterilization (Smith and Baird-Parker, 1964). To prepare a stock solution of sulphamezathine, dissolve 0.5 g of pure sodium sulphamezathine in 25 ml of 1/10 *N* sodium hydroxide and make up to 250 ml with distilled water; 27.5 ml of this solution delivers 0.055 g of sulphamezathine.

Directions for preparation of complete medium   To melted and tempered

1/Acknowledgment is made of frequent use of the compendia of media supplied by Difco Laboratories Inc. (1953, 1962), Baltimore Biological Laboratory, Inc. (1956), and Oxoid Division, Oxo Ltd. (1965).
2/Difco B123 was used in development studies.
3/Oxoid L30 was used in development studies.
4/Difco B127 was used in development studies.

(45–50° C) 90 ml portions of basal medium add aseptically the following prewarmed (45–50° C) filtered-sterilized solutions in the amounts designated:

(i)	20% w/v solution of glycine	6.3 ml
(ii)	1% w/v solution of potassium tellurite[5]	1.0 ml
(iii)	20% w/v solution of sodium pyruvate	5.0 ml
(iv)	Oxoid new improved egg-yolk emulsion	5.0 ml
	(to prepare egg emulsion in the laboratory, if preferred, use the method described by Billing and Luckhurst, 1957)	

Mix well and pour immediately in 15 ml amounts into Petri dishes.

Notes on storage The basal medium can be stored for several months at room temperature in screw-capped bottles, without loss of selectivity or increased toxicity to *Staphylococcus aureus*. Stock solutions of 20% glycine and 1% tellurite can be stored for several months at room temperature. The sodium pyruvate solution is best stored at 5° C and replaced monthly. Poured plates of the complete medium cannot be stored satisfactorily either at room temperature or at 5° C and should therefore be freshly prepared and used preferably within 24 hours of pouring and not after more than 48 hours of storage.

Dehydrated Baird-Parker Agar can be used; prepare as directed on the container. (*Note* Most workers have found that the commercial form of the medium is not as satisfactory as that made in the laboratory. However, the dehydrated product may be used where facilities are not available for preparing the medium directly. The commercial medium tends to be more inhibitory than the laboratory-made version and some strains form very small colonies; the egg-yolk reaction is comparable for the two media.)

2 BISMUTH SULPHITE AGAR (Wilson and Blair, modified)

Formula

Beef extract	5.0	g
Peptone or polypeptone	10.0	g
Glucose	5.0	g
Sodium monohydrogen phosphate	4.0	g
Ferrous sulphate	0.3	g
Bismuth sulphite indicator	8.0	g
Brilliant green	0.025	g
Agar	20.0	g

5/The source is critical. The supply for development studies was obtained from British Drug House, Ltd.

Directions Suspend ingredients in 1 litre of distilled water, mix well, and heat to boiling with frequent agitation to dissolve soluble materials. A precipitate is formed which will not dissolve. Cool to about 45–50° C and pour in 15–20 ml quantities into Petri dishes. Do not autoclave. The selectivity of the medium decreases 48 hours after the time of preparation.

Dehydrated Bismuth Sulphite Agar is available commercially; prepare as directed on the container.

3 BISMUTH SULPHITE AGAR (McCoy, 1962)

Formula for basal medium

Peptone	10	g
Yeast extract	0.8	g
Ferric citrate	0.4	g
Brilliant green	0.01	g

Directions for complete medium Add ingredients to 1 litre of distilled water, heat to boiling with agitation to obtain complete solution, cool to 50–55° C, adjust pH to about 7.5, dispense in 500 ml volumes in flasks or bottles, and sterilize by autoclaving at 116° C for 20 minutes. Cool to 55° C and add 100 ml of bismuth sulphite indicator.

BISMUTH SULPHITE INDICATOR

Formula

Bismuth ammonia-citrate	3	g
Sodium sulphite (anhyd.)	10	g
Sodium monohydrogen phosphate (anhyd.)	5	g
Dextrose	5	g

Directions Dissolve ingredients in 100 ml of distilled water, heat to boiling, cool to room temperature, and add to the basal medium as described above.

4 BISMUTH SULPHITE PHENOL-RED AGAR

Formula for basal medium

Beef extract	5	g
Peptone	10	g
Sodium chloride	20	g
Phenol red	0.02	g
Agar	15	g

Directions for complete medium Add above ingredients to 1 litre of distilled water, heat to boiling with agitation to obtain complete solution, cool to 45–50° C, adjust pH to 8.6 ± 0.2, and sterilize by autoclaving at 121° C for 15 minutes. Cool to 45–50° C and add 100 ml of bismuth sulphite solution and 1 ml of 95% ethyl alcohol.

Bismuth sulphite solution Dissolve (*a*) 100 g of sodium sulphite in 500 ml of boiling water, (*b*) 30 g of ammonium bismuth citrate in 250 ml of boiling water, (*c*) 50 g of sucrose, 5 g of mannitol, and 15 g of sodium bicarbonate in 300 ml of boiling water. Mix solutions (*a*) and (*b*) together and boil for a further 1 minute. Add solution (*c*) to the mixture while hot and mix. Since the final preparation contains a white precipitate, mix thoroughly before using. The reagent remains usable for 1 month if stored in the refrigerator.

5 BISMUTH SULPHITE SALT BROTH

Formula for basal medium

Peptone	10	g
Sodium chloride	25	g
Potassium chloride	0.7	g
Magnesium chloride hexahydrate	5	g

Directions for complete medium Dissolve above ingredients in 950 ml of distilled water and adjust pH to 9.1 by addition of 10% aqueous solution of Na_2CO_3. Sterilize by autoclaving at 121° C for 15 minutes, cool to room temperature, and add aseptically 100 ml of bismuth sulphite solution and 1 ml of 95% ethyl alcohol. Mix well and dispense with aseptic precaution in 10 ml portions into sterile culture tubes. Do not heat.

Bismuth sulphite solution Dissolve (*a*) 20 g of sodium sulphite in 100 ml of boiling water, (*b*) 0.1 g of ammonium bismuth citrate in 100 ml of boiling water, and (*c*) 20 g of mannitol in 100 ml of boiling water. Mix solutions (a) and (*b*), boil the mixture for 1 minute, and add solution (*c*). The final mixture is white in colour and turbid; mix well before using. The reagent remains usable for 1 month if stored in the refrigerator.

Dehydrated Bismuth Sulphite Salt Broth is available from "Eiken" Ltd. (Nihon Eiyo-Kagaku Co. Ltd., Hongo 1–33, Bunkyo-ku, Tokyo, Japan) and "Nissui" Ltd. (Nissui Seiyaku Co. Ltd., Sendagi 3–22, Bunkyo-ku, Tokyo, Japan). Reconstitute as directed on the container.

6 BLOOD AGAR

Basal medium (Dolman, 1957)

Beef extract	3 g
Peptone	5 g
Sodium chloride	8 g
Agar	15 g

Directions for complete medium   Add ingredients to 500 ml of distilled water and heat to boiling with frequent agitation to obtain complete solution. Cool to 50–60° C and adjust reaction so that pH after autoclaving will be 7.3 ± 0.2. Dispense in 95 ml volumes and autoclave at 121° C for 15 minutes. Before use, melt agar, cool to 50° C, and add 5 ml of sterile citrated human or rabbit blood to each 95 ml of basal medium. Mix well and pour in 15 ml volumes into Petri dishes. Dry surface of plates before inoculation (for note on drying plates, see step 9 in the introduction to this chapter).

7 BRAIN HEART INFUSION AGAR

Formula

Calf brains, infusion from	200	g
Beef heart, infusion from	250	g
Peptone or proteose peptone	10	g
Sodium chloride	5	g
Sodium monohydrogen phosphate	2.5	g
Glucose	2	g
Agar	15	g

Directions   Add ingredients to 1 litre of distilled water, heat to boiling with agitation to obtain complete solution, and sterilize by autoclaving at 121° C for 15 minutes.

Dehydrated preparations of the medium are available commercially; prepare as directed on the container.

8 BRAIN HEART INFUSION BROTH
Formula

Calf brains, infusion from	200	g
Beef heart, infusion from	250	g

Peptone or proteose peptone	10	g
Sodium chloride	5	g
Sodium monohydrogen phosphate	2.5	g
Glucose	2	g

Directions Dissolve ingredients in 1 litre of distilled water, dispense in 5 ml volumes in tubes, and sterilize by autoclaving at 121° C for 15 minutes. Final pH, 7.4. For best results use medium on the day it is prepared; otherwise boil or steam it a few minutes and then cool before use.

Dehydrated preparations of the medium are available commercially; reconstitute as directed on the container.

9 BRILLIANT GREEN AGAR

Formula

Yeast extract	3.0	g
Proteose peptone or polypeptone	10.0	g
Sodium chloride	5.0	g
Lactose	10.0	g
Saccharose	10.0	g
Phenol red	0.08	g
Brilliant green	0.0125	g
Agar	20.0	g

Directions Suspend ingredients in 1 litre of distilled water, and heat to boiling with frequent agitation to obtain complete solution. Cool to 50–60° C, adjust reaction so that the final pH will be 6.9 ± 0.1, and dispense as required in flasks. Autoclave at 121° C for 12 minutes (additional heating decreases selectivity and less heating increases selectivity).

The medium is available commercially in dehydrated form; prepare as directed on the container.

10 BRILLIANT-GREEN LACTOSE BILE BROTH 2%

Formula

Peptone	10	g
Lactose	10	g
Ox-gall	20	g
Brilliant green	0.0133	g

Directions   Dissolve the peptone and lactose in 500 ml of water and add the ox-gall dissolved in 200 ml of water. Bring the volume to approximately 975 ml with water and adjust the pH to 7.4. Add 13.3 ml of a 1% aqueous solution of brilliant green, bring the total volume to 1 litre, stir, and filter through cotton if necessary. Dispense in 10 ml volumes into 150 × 15 mm tubes containing inverted Durham fermentation vials (75 × 10 mm) and sterilize by autoclaving at 121° C for 10 minutes.

The medium is available commercially in dehydrated form; prepare as directed on the container.

11 BRILLIANT-GREEN MACCONKEY AGAR

Formula

Peptone	20.0	g
Lactose	10.0	g
Bile salts	5.0	g
Sodium chloride	5.0	g
Neutral red	0.075	g
Brilliant green	0.0125	g
Agar	15.0	g

Directions   Suspend ingredients in 1 litre of distilled water, heat to boiling with frequent agitation to obtain complete solution, cool to 50–60° C, adjust reaction so that the pH after autoclaving will be 7.4 ± 0.1, dispense as required in flasks, and autoclave at 121° C for 12 minutes.

MacConkey agar is available commercially in dehydrated form; prepare as directed on the container, adding brilliant green to reconstituted powder.

12 BRILLIANT-GREEN SULPHADIAZINE AGAR

Formula

The ingredients for Medium 9 at the same concentrations, plus
Sulphadiazine	0.08 g

Directions   Prepare as directed for Medium 9, adding the sulphadiazine aseptically (preferably as 10 ml of a 0.8% aqueous solution), after sterilization, to tempered agar (50° C) immediately before pouring plates (Galton *et al.*, 1954).

13 BTP TEEPOL AGAR

Formula

Beef extract	5	g
Peptone	10	g
Sodium chloride	30	g
Sucrose	10	g
Teepol[6]	2	ml
Bromthymol blue	0.08	g
Agar	15	g

Directions   Add all ingredients to 1 litre of distilled water, heat to boiling with agitation to obtain complete solution, cool to 50–55° C, adjust pH to 7.8 ± 0.2, and pour in 15–20 ml amounts into Petri dishes.

Dehydrated BTP Teepol Agar is available from "Eiken" Ltd. and "Nissui" Ltd. (see Medium 5 for complete names and addresses of these companies); reconstitute as directed on the container.

14 BUFFERED GLUCOSE BROTH

Formula

Proteose peptone	5 g
Glucose	5 g
Potassium monohydrogen phosphate	5 g

Directions   Add ingredients to 1 litre of distilled water, heat gently with stirring to obtain complete solution, distribute in 5 ml volumes in culture tubes, and sterilize by autoclaving at 121° C for 15 minutes.

15 BUFFERED GLUCOSE SALT BROTH

Formula

Peptone	7 g
Glucose	5 g
Potassium monohydrogen phosphate	5 g
Sodium chloride	30 g

Directions   Dissolve ingredients in 1 litre of distilled water, dispense in 5 ml volumes into culture tubes, and autoclave at 121° C for 15 minutes.

6/Mixture of sodium salts of sulphated fatty alcohols, available from Shell Oil Co.

16 COOKED MEAT ENRICHMENT MEDIUM

Formula

Beef heart or liver (finely chopped, treated and dried as described in Medium 17)	454 g
Peptone or proteose peptone	20 g
Glucose	2 g
Sodium chloride	5 g

Directions Add ingredients to 1 litre of distilled water and agitate with heating to dissolve the salt, sugar, and peptone. Cool and adjust reaction so that the pH after autoclaving will be approximately 7.0. Agitating flask frequently to keep meat pieces in suspension, distribute medium in 10 ml volumes in 150 × 15 mm screw-capped tubes. Sterilize tubes at 121° C for 15 minutes. If the medium is not used the same day it is sterilized, place tubes in a flowing steam or boiling water bath for a few minutes to drive off dissolved gases, and allow to cool without agitation.

Cooked Meat Enrichment Medium is available commercially in dehydrated form; prepare as directed on the container.

17 COOKED MEAT MEDIUM

Directions Heat 1 litre of 1/20 N aqueous sodium hydroxide to boiling and add 1000 g of minced fat-free ox heart. Mix thoroughly, bring to boiling point, and allow to simmer for 20 minutes, stirring frequently. The reaction of the mixture should be about pH 7.5. Strain through several layers of muslin cloth, squeeze out excess liquid, and spread partially dried meat on filter paper to dry further. Place dried meat in test tubes (15 × 150 mm approx.) to a depth of 2.5 cm, and add sufficient Nutrient Broth No. 2 (Medium 47) to give a total depth in the tubes of about 4 cm. Sterilize by autoclaving at 121° C for 20 minutes.

Infusion from meat as prepared above, supplemented with sodium chloride (0.5%), sodium monohydrogen phosphate (0.08%), and peptone (1%) may be used in place of Nutrient Broth No. 2 (Dolman, 1957).

18 DESOXYCHOLATE CITRATE AGAR (Hynes, 1942)

Formula

Beef extract	5.0 g
Proteose peptone	5.0 g
Lactose	10.0 g

Sodium citrate	8.5	g
Sodium thiosulphate	5.4	g
Ferric citrate	1.0	g
Sodium desoxycholate	5.0	g
Neutral red	0.02	g
Agar	12.0	g

Directions Suspend ingredients in 1 litre of distilled water, mix well, and heat to boiling with agitation to obtain complete solution. Cool to 45–50° C and pour in 15–20 ml amounts into Petri dishes. Final pH, 7.3 approximately. This medium does not require sterilization by auto-claving and should not be remelted.

The medium is available commercially in dehydrated form; prepare as directed on the container.

19 E.C. BROTH

Formula

Tryptose or trypticase	20	g
Lactose	5	g
Bile Salts No. 3	1.5	g
Potassium monohydrogen phosphate	4	g
Potassium dihydrogen phosphate	1.5	g
Sodium chloride	5	g

Directions Dissolve ingredients in 1 litre of distilled water, heating gently, if necessary, to obtain complete solution. Distribute in 10 ml volumes in 150 × 15 mm tubes containing inverted Durham fermentation vials (75 × 10 mm). Autoclave at 121° C for 10 minutes. Final pH, 6.8.

Dehydrated E.C. Broth is available commercially; reconstitute as directed on the container.

20 EGG-YOLK AZIDE AGAR

Formula for basal medium

Beef extract	5.5	g
Peptone	10	g
Sodium chloride	3	g
Sodium monohydrogen phosphate	0.2	g
Agar	15	g
Sodium azide	0.15	g

Directions for preparation of complete medium   Dissolve all ingredients except sodium azide in 1 litre of distilled water, adjust pH to 7.6, and autoclave at 121° C for 30 minutes. Cool to 50–60° C, add the 0.15 g of sodium azide, mix, and resterilize at 120° C for 30 minutes. Cool to 50° C and add 150 ml of sterile egg-yolk–saline solution. Mix well and pour immediately in 15 ml quantities into Petri dishes.

Egg-yolk–saline solution   To prepare, mix 1 part of sterile commercially available egg-yolk emulsion with 1 part of sterile saline (0.85% aqueous solution of sodium chloride). To prepare egg-yolk emulsion in laboratory, if preferred, use method described by Billing and Luckhurst (1957) or the method outlined in Medium 35. If dehydrated egg yolk is used, check prepared medium for production of typical reaction with a coagulase-positive strain of *Staphylococcus*. Some commercial brands are not satisfactory.

21 ENDO AGAR

Formula

Peptone	10	g
Lactose	10	g
Potassium monohydrogen phosphate	3.5	g
Sodium sulphite	2.5	g
Basic fuchsin	0.4	g
Agar	15	g

Directions   Medium should be used the same day it is prepared. Add all ingredients except the sodium sulphite and basic fuchsin to 1 litre of distilled water, heat to boiling until solution is complete, dispense in 100 ml quantities, and autoclave at 121° C for 15 minutes. Immediately before use, melt the agar and to each 100 ml quantity add 1 ml of a 4% basic fuchsin solution (in 95% ethyl alcohol) and 2.5 ml of a 10% aqueous solution of sodium sulphite. (The sulphite solution must be prepared immediately before use.) Mix thoroughly and pour in 15–20 ml amounts into Petri plates. Final pH, 7.5.

Dehydrated Endo Agar is available commercially; reconstitute as directed on the container.

22 ENTEROBACTERIACEAE ENRICHMENT BROTH

Formula

Peptone	10	g
Glucose	5	g
Sodium monohydrogen phosphate dihydrate	8	g

Potassium dihydrogen phosphate	2	g
Ox-gall	20	g
Brilliant green	0.015	g

Directions   Only purified Ox-gall and Brilliant green should be used so as not to inhibit very low numbers of debilitated cells of the Enterobacteriaceae. Dissolve ingredients in 1 litre of distilled water, dispense in 12 ml volumes into tubes and in 100 ml volumes into flasks, and heat to boiling.

This medium is available commercially in dehydrated form; reconstitute as directed on the container.

23 EOSIN METHYLENE-BLUE (EMB) AGAR

Formula

Peptone	10	g
Lactose	10	g
Sucrose	5	g
Potassium monohydrogen phosphate	2	g
Eosin Y	0.4	g
Methylene blue	0.063	g
Agar	15	g

Directions   Add all ingredients except dyes to 955 ml of distilled water. Add 20 ml of 2% aqueous solution of eosin Y and 25 ml of 0.25% aqueous solution of methylene blue. Heat to boiling to obtain complete solution, cool to 50–60° C, mix well, dispense as required (usually in 100–200 ml volumes), and autoclave at 121° C for 15 minutes.

Dehydrated EMB Agar is available commercially; reconstitute as directed on the container.

24 FLUID THIOGLYCOLLATE MEDIUM

Formula

Trypticase or casitone	15	g
l-Cystine	0.5	g
Glucose	5	g
Yeast extract	5	g
Sodium chloride	2.5	g
Sodium thioglycollate	0.5	g
Resazurin	0.001	g
Agar	0.75	g

Directions Add ingredients to 1 litre of distilled water, mix well, and heat to boiling to obtain complete solution. Cool to 50–60° C, adjust reaction so that pH after autoclaving will be 7.1 ± 1, and dispense in 10 ml volumes in 150 × 15 mm screw-capped tubes, or in 25 ml volumes in 250 × 25 mm screw-capped tubes. Add approximately 0.1 g of calcium carbonate to tubes before addition of medium. Autoclave at 121° C for 15 minutes and cool quickly. Immediately before use, heat tubes in flowing steam for 10 minutes to drive off dissolved oxygen and cool rapidly in tap water.

Fluid Thioglycollate Medium is available commercially in dehydrated form; prepare as directed on the container.

25 GILLIES MEDIUM 1 (Gillies, 1956)

Formula

Proteose peptone	15	g
Yeast extract	2	g
Beef extract	2	g
Glucose	1	g
Mannitol	10	g
Bromthymol blue	0.025	g
Cresol red	0.008	g
Thymol blue	0.020	g
Agar	16	g
Urea	10	g

Directions Add all ingredients except urea to 1 litre of distilled water, mix well, and heat to boiling with frequent agitation to dissolve completely. Cool to 50–60° C and adjust reaction so that pH after autoclaving will be 7.2 ± 0.1. Autoclave in flask at 121° C for 15 minutes, cool to 60° C, add 50 ml of a filter-sterilized 20% aqueous solution of urea, and agitate thoroughly to obtain a homogeneous solution. Dispense aseptically into sterile tubes in quantities sufficient to give a 25–30 mm butt and a generous slant when cooled in a slanted position.

Gillies Medium 1 (lacking only the urea) is available commercially in dehydrated form; prepare as directed on the container.

26 GILLIES MEDIUM 2 (Gillies, 1956)

Formula

Peptone	10	g

Tryptone	10	g
Sucrose	10	g
Salicin	10	g
Sodium chloride	5	g
Sodium monohydrogen phosphate (anhyd.)	0.14	g
Sodium thiosulphate	0.025	g
Bromthymol blue	0.010	g
Agar	3.0	g

Directions   Add ingredients to 1 litre of distilled water, mix well, and heat to boiling with frequent agitation to dissolve completely. Dispense into tubes to give a depth of 50–60 mm, and autoclave at 121° C for 15 minutes. Final pH, 7.4 ± 0.1. After tubes are inoculated, insert into the top of each a strip of lead acetate paper (Reagent 4) and a strip of indole paper (Reagent 2). Hold the strips above the medium by means of the cotton plug and do not permit the two papers to touch each other.

Gillies Medium 2 is available commercially in dehydrated form; prepare as directed on the container.

27 GLUCOSE BROTH

Formula

Beef extract	3 g
Peptone	5 g
Glucose	10 g

Directions   Dissolve ingredients in 1 litre of distilled water, adjust reaction for a post-sterilization pH of 7.3 ± 0.2, dispense in 7–10 ml volumes into tubes, and sterilize by autoclaving at 121° C for 15 minutes.

28 GLUCOSE SALT MEDIUM

Formula

Tryptone	2	g
Glucose	10	g
Sodium chloride	5	g
Potassium monohydrogen phosphate	0.3	g
Agar	3	g
Bromthymol blue (0.1% alcoholic solution)	30	ml

Directions   Add all ingredients except glucose to 870 ml of distilled water, heat to boiling with agitation until solution is complete, cool, dispense in 4.5–5 ml volumes into culture tubes, autoclave (121° C for 15 minutes), and cool to 45° C. Sterilize a 10% aqueous solution of glucose by filtration (see step 8 of the introduction to this chapter) and aseptically add 0.5 ml to each tube of the base. Mix well and cool to room temperature. Final pH, 7.1.

29 H BROTH

Formula

Peptone	5	g
Tryptone	5	g
Beef extract	3	g
Glucose	1	g
Sodium chloride	5	g
Potassium monohydrogen phosphate	2.5	g

Directions   Dissolve ingredients in 1 litre of distilled water, distribute in 4 ml amounts in small tubes (e.g., 100 × 13 mm), and sterilize by autoclaving at 115° C for 15 minutes. Final pH, 7.2.

H Broth is available commercially in dehydrated form; reconstitute as directed on the container.

30 HORSE-BLOOD AGAR

Agar base, formula 1

Beef extract	10	g
Peptone	10	g
Sodium chloride	5	g
Agar	15	g

Agar base, formula 2

Beef heart, infusion from	500	g
Tryptone or thiotone	10	g
Sodium chloride	5	g
Agar	15	g

Directions for complete medium   Both media are prepared the same way. Suspend ingredients in 1 litre of distilled water, mix well, and heat to boiling with frequent agitation to obtain complete solution. Cool to 50–60° C and, if necessary, adjust reaction so that pH after autoclaving will be 7.2 ± 0.2. Dispense as required and autoclave at 121° C for 15 minutes.

Addition of blood Cool autoclaved basal medium to 45° C and add 5–7 ml of sterile defibrinated horse blood per 100 ml of medium. Mix thoroughly and pour in 15–20 ml amounts into Petri dishes already containing 10 ml of solidified basal medium. Dry surfaces of plates before use (see step 9 of the introduction to this chapter). Store plates at 5–8° C until required.

Both agar bases are available commercially in dehydrated form; prepare as directed on the container. Defibrinated horse blood is also available commercially.

31 HUGH-LEIFSON SALT MEDIUM (Hugh and Leifson, 1953)

Formula

Peptone	2.7	g
Glucose	10	g
Sodium chloride	30	g
Potassium monohydrogen phosphate	0.3	g
Bromthymol blue	0.05	g
Agar	3	g

Directions Add ingredients to 1 litre of distilled water, heat to boiling to obtain complete solution, cool to 50° C, adjust pH to 7.2 ± 0.1, distribute in 7–10 ml volumes in culture tubes, and autoclave at 121° C for 15 minutes.

Dehydrated Hugh-Leifson Medium is available from "Nissui" Ltd. and "Eiken" Ltd. (see Medium 5 for complete names and addresses of these companies); reconstitute as directed on the container and add sodium chloride to give a final concentration of 3%.

32 JORDAN SALT AGAR (Phenol-Red Tartrate Agar; Jordan and Harmon, 1928)

Formula

Peptone	10	g
Sodium potassium tartrate	10	g
Sodium chloride	30	g
Phenol red	0.08	g
Agar	12	g

Directions Add ingredients to 1 litre of distilled water, heat to boiling to obtain complete solution, dispense into culture tubes to a depth of 2–3 cm, and autoclave at 121° C for 15 minutes.

Dehydrated Jordan Agar is available commercially; reconstitute as directed on the container and add sodium chloride to give a final concentration of 3%.

33 KOSER CITRATE BROTH (Koser, 1923)

Formula

Sodium ammonium phosphate	1.5	g
Potassium dihydrogen phosphate	1	g
Magnesium sulphate heptahydrate	0.2	g
Sodium citrate dihydrate	2.5	g
Bromthymol blue	0.016	g

Directions Dissolve all ingredients except bromthymol blue in 990 ml of distilled water. Add 10 ml of a filtered 0.16% aqueous solution of bromthymol blue, mix, and distribute in 5 ml volumes in culture tubes. Sterilize by autoclaving at 121° C for 15 minutes. Final pH, 6.8 (approx.).

Dehydrated Koser Citrate Broth is available commercially; reconstitute as directed on the container.

34 LACTOSE BROTH

Formula

Beef extract	3	g
Peptone or polypeptone	5	g
Lactose	5	g

Directions Dissolve ingredients in 1 litre of distilled water, adjust reaction so that final pH will be 6.7–6.9, dispense as required, and sterilize at 121° C for 15 minutes.

Dehydrated Lactose Broth is available commercially; prepare as directed on the container.

35 LACTOSE EGG-YOLK MILK AGAR (Willis and Hobbs, 1959)

Formula

Lactose	12	g
Neutral red (1% aqueous solution)	3.25	ml
Agar	12	g
Meat infusion broth (pH 7.0)	1000	ml

Directions Add lactose, neutral red, and agar to the infusion broth, autoclave mixture at 121° C for 20 minutes, cool to 50–55° C, and add with aseptic precaution 37.5 ml of egg-yolk suspension and 150 ml of stock milk. Sodium thioglycollate (0.1% final concentration) may be added with the egg-yolk suspension to assist the growth of the stricter anaerobes. Neomycin sulphate (250 μg per ml) can also be added to reduce the growth of facultative anaerobes. Mix well and pour in 15–20 ml amounts into Petri dishes.

Egg-yolk suspension Scrub and sterilize shells by immersing fresh eggs in 0.1% aqueous solution of mercury chloride. Aseptically remove yolks from eggs and place in sterile graduated cylinder. Add an equal volume of sterile saline (0.85% sodium chloride) and mix well.

Stock milk Centrifuge ordinary whole milk (10 minutes at 2000–3000 rpm in a clinical centrifuge is sufficient) and autoclave fat-free portion at 121° C for 15 minutes.

36 LAURYL SULPHATE TRYPTOSE BROTH

Formula

Tryptose, tryptone, or trypticase	20	g
Lactose	5	g
Potassium monohydrogen phosphate	2.75	g
Potassium dihydrogen phosphate	2.75	g
Sodium chloride	5	g
Sodium lauryl sulphate	0.1	g

Directions Dissolve ingredients in 1 litre of distilled water, dispense in 10 ml volumes into 150 × 15 mm tubes containing inverted Durham fermentation vials (75 × 10 mm), sterilize by autoclaving at 121° C for 10 minutes. Final pH, 6.8 (approx.).

The medium is available commercially in dehydrated form; prepare as directed on the container.

37 LITMUS MILK

Formula

Skim milk powder	100	g
Litmus	5	g

Directions Place ingredients in a 2-litre flask and add 1 litre of distilled water gradually, agitating continually. Adjust pH to 6.8. Strain through muslin or cotton, distribute in 10 ml amounts in 150 × 15 mm tubes, and autoclave at 121° C for 5 minutes.

It should be noted that, when hot, the medium becomes colourless, but that on cooling the colour returns.

Litmus milk is available commercially in dehydrated form; prepare as directed on the container.

38 LIVER VEAL EGG-YOLK AGAR

Formula for basal medium

Liver, infusion from	50	g
Veal, infusion from	500	g
Proteose peptone	20	g
Neopeptone	1.3	g
Tryptone	1.3	g
Dextrose	5	g
Soluble starch	10	g
Isoelectric casein	2	g
Sodium chloride	5	g
Sodium nitrate	2	g
Gelatin	20	g
Agar	15	g

Directions for complete medium   Add ingredients to 1 litre of distilled water, mix well, and heat to boiling with frequent agitation to obtain complete solution. Cool to 50–60° C, adjust reaction so that pH after autoclaving will be 7.3 ± 1, dispense in 90 ml volumes into bottles, and sterilize at 121° C for 15 minutes. Immediately before use melt in boiling water or steamer, cool to 50° C, and add 10 ml of freshly prepared egg-yolk suspension (prepared as described in Medium 35) or 10 ml of a suitable commercially available egg-yolk preparation.

Liver Veal Egg-Yolk Agar is available commercially in dehydrated form; prepare as directed on the container.

39 LYSINE IRON AGAR (Edwards and Fife, 1961)

Formula

Peptone or gelysate	5	g
Yeast extract	3	g
Glucose	1	g
l-Lysine	10	g
Ferric ammonium citrate	0.5	g
Sodium thiosulphate	0.04	g

Bromcresol purple	0.02	g
Agar	15	g

Directions   Suspend ingredients in 1 litre of distilled water, mix well, heat to boiling with frequent agitation to dissolve completely, cool to 50–60° C, and adjust reaction for post-sterilization pH of 6.7 ± 0.1. Dispense in 4 ml volumes into small tubes (about 13 × 100 mm) and cap or plug so that aerobic conditions are maintained during use. Autoclave at 121° C for 12 minutes. Allow tubes to cool in slanted position to give 3 cm butts and 2 cm slants.

Lysine Iron Agar is available commercially in dehydrated form; prepare as directed on the container.

40 MAC CONKEY AGAR

Formula

Peptone or gelysate	17	g
Proteose peptone or polypeptone	3	g
Lactose	10	g
Bile salts or bile salts No. 3	1.5	g
Sodium chloride	5	g
Agar	13.5	g
Neutral red	0.03	g
Crystal violet	0.001	g

Directions   Suspend ingredients in 1 litre of distilled water, mix well, and heat to boiling with frequent agitation to obtain complete solution. Cool to 50–60° C and adjust reaction so that pH after autoclaving will be 7 ± 0.1. Dispense into flasks or tubes as required and autoclave at 121° C for 15 minutes.

MacConkey Agar is available commercially in dehydrated form; prepare as directed on the container.

41 MAC CONKEY BROTH

Formula

Peptone	20	g
Lactose	10	g
Bile salts	5	g
Sodium chloride	5	g
Neutral red or	0.075	g
bromcresol purple	0.01	g

Directions Dissolve ingredients in 1 litre of distilled water, adjust pH to 7.6, dispense in 10 ml volumes into 150 × 15 mm tubes containing inverted Durham fermentation vials, and sterilize by autoclaving at 121° C for 15 minutes.

Dehydrated MacConkey Broth is available commercially; reconstitute as directed on the container.

42 MCCLUNG-TOABE EGG-YOLK AGAR (McClung and Toabe, 1947)

Formula for basal medium

Proteose peptone	40	g
Sodium monohydrogen phosphate septahydrate	5	g
Potassium dihydrogen phosphate	1	g
Sodium chloride	2	g
Magnesium sulphate	0.1	g
Glucose	2	g
Agar	25	g

Directions for complete medium Add ingredients to 1 litre of distilled water, mix well, and heat to boiling to obtain complete solution. Cool to 50–60° C, adjust reaction so that pH after autoclaving will be 7.6 ± 1, dispense in 90 ml volumes, and sterilize at 121° C for 20 minutes. Before use, melt agar, cool to 50° C, add 10 ml of egg-yolk suspension (prepare as described in Medium 35 above) to each 90 ml of basal medium, mix well, and pour in 15 ml quantities into Petri dishes. Dry surface of plates before use (see step 9 of introduction to this chapter for note on drying plates).

43 MILK SALT AGAR

Formula for basal medium

Beef extract	3	g
Peptone	5	g
Sodium chloride	65	g
Agar	15	g

Directions for complete medium Suspend ingredients in 1 litre of distilled water, mix, and heat to boiling with frequent agitation to obtain complete solution. Cool to 50–60° C, adjust reaction so that pH after autoclaving will be 7.4 ± 0.1, distribute in 100 ml volumes in flasks or bottles, and sterilize at 121° C for 15 minutes. Cool to 50° C and add 10 ml of sterile skim milk (110° C for 15 minutes) to each 100 ml of

base. Mix well and pour into plates. The plates can be stored at 5–8° C
for 2–3 days, if necessary.

44 MOTILITY-NITRATE MEDIUM

Formula

Beef extract	3 g
Peptone	5 g
Potassium nitrate	1 g
Agar	3 g

Directions Add ingredients to 1 litre of distilled water, mix well, and
heat to boiling to obtain complete solution. Cool to 50–60° C, adjust
reaction so that the pH after autoclaving will be 7.0 ± 0.2, dispense in
10 ml volumes into 150 × 15 mm screw-capped tubes, and sterilize at
121° C for 15 minutes.

45 NITRATE BROTH

Formula

Tryptone or trypticase	20 g
Sodium monohydrogen phosphate	2 g
Glucose	1 g
Agar	1 g
Potassium nitrate	1 g

Directions Add ingredients to 1 litre of distilled water, heat to boiling
with agitation until solution is complete, dispense in 5 ml volumes in culture tubes, and sterilize at 121° C for 15 minutes. If prepared medium is
more than 2 days old, it should be boiled about 2 minutes before use.

Nitrate Broth is available commercially in dehydrated form; reconstitute
as directed on the container.

46 NUTRIENT AGAR

Formula

Beef extract	3 g
Peptone	5 g
Agar	15 g

Directions Add ingredients to 1 litre of distilled water, heat to boiling
until solution is complete, cool to 50–60° C, and adjust reaction so that
the pH after sterilization will be 6.8–7.0. Distribute in tubes for slants or
in bulk for plates as required and autoclave at 121° C for 15 minutes.

Dehydrated Nutrient Agar is available commercially; reconstitute as directed on the container.

47 NUTRIENT BROTH NO. 2

Formula

Beef extract	10 g
Peptone	10 g
Sodium chloride	5 g

Directions Dissolve ingredients in 1 litre of distilled water, adjust reaction so that the pH after autoclaving will be 7.5 ± 0.1, dispense as required (see Medium 17), and autoclave at 121° C for 15 minutes.

Nutrient Broth No. 2 is available commercially in dehydrated form; prepare as directed on the container.

48 NUTRIENT GELATIN

Formula

Beef extract	3 g
Peptone	5 g
Gelatin	120 g

Directions Suspend ingredients in 1 litre of distilled water, mix well, and heat to boiling to obtain complete solution. Adjust reaction, if necessary, so that the pH after autoclaving will be 6.9 ± 0.1. Dispense into tubes to a depth of 2–3 cm and sterilize at 121° C for 15 minutes.

Dehydrated Nutrient Gelatin is available commercially; prepare as directed on the container.

49 NUTRIENT SALT AGAR

Formula

Beef extract	3 g
Peptone	5 g
Sodium chloride	30 g
Agar	15 g

Directions Add ingredients to 1 litre of distilled water, heat to boiling with agitation to obtain complete solution, distribute in 7–10 ml amounts in tubes, autoclave for 15 minutes at 121° C, and allow tubes to cool in the sloped position. Final pH, approximately 7.

Dehydrated Nutrient Agar is available commercially; reconstitute as directed on the container and add sodium chloride to give a final concentration of 3%.

50 NUTRIENT SALT GELATIN

Formula

Beef extract	3 g
Peptone	5 g
Sodium chloride	30 g
Gelatin	120 g

Directions Add ingredients to 1 litre of distilled water, heat to boiling with agitation to obtain complete solution, cool to 50° C, dispense into culture tubes to a depth of 2–3 cm, and autoclave at 121° C for 15 minutes.

Dehydrated Nutrient Gelatin is available commercially; reconstitute as described on the container and add sodium chloride to give a final concentration of 3%.

51 PACKER'S CRYSTAL-VIOLET AZIDE BLOOD AGAR (Packer, 1943)

Formula

Tryptose	15	g
Beef extract	3	g
Sodium chloride	5	g
Agar	30	g
Defibrinated sheep blood	50	ml
Crystal violet	0.002	g
Sodium azide	0.5	g

Directions Add the tryptose, beef extract, sodium chloride, and agar to 1 litre of distilled water and heat to boiling to obtain complete solution. Cool to 50–60° C and adjust reaction so that the pH after sterilization will be 7.0–7.2. Distribute in 100 ml volumes in bottles or flasks, sterilize by autoclaving (121° C, for 15 minutes), cool to 50° C, and add the following to each 100 ml:

(a) 5 ml of fresh defibrinated sheep blood (store blood no longer than 1 week before use).

(b) 0.4 ml of a sterile 0.05% aqueous solution of crystal violet (sterilize crystal violet solution at 121° C for 20 minutes and store at 1–5° C).

(c) 1 ml of a sterile 5% aqueous solution of sodium azide (sterilize sodium azide solution by filtration).

Mix well, temper at 45 ± 1° C, and pour plates as required.

52 PEPTONE BROTH

Directions   Dissolve 10 g of peptone in 1 litre of distilled water, adjust pH to 7.2 ± 0.2, distribute in 10 ml portions in culture tubes, and autoclave at 121° C for 15 minutes.

53 PEPTONE DILUTION FLUID (Straka and Stokes, 1957)

Directions   Dissolve 1 g of peptone in 1 litre of distilled water and adjust pH to 7.0 ± 0.1. Fill dilution bottles or tubes with predetermined volume so that after autoclaving (121° C for 15 minutes) the volume will be ±2% of that desired; or, if containers are calibrated, aseptically readjust volume by pipette after autoclaving.

54 PEPTONE SALT (3%) BROTH

Formula

Peptone	10 g
Sodium chloride	30 g

Directions   Add ingredients to 1 litre of distilled water, adjust pH to 7.2 ± 0.2, distribute in culture tubes to a depth of 2–3 cm, and autoclave at 121° C for 15 minutes.

55 PEPTONE SALT (4%) BROTH

Formula

Peptone	10 g
Sodium chloride	40 g

Directions   Prepare as described for Medium 54.

56 PEPTONE SALT (7%) BROTH

Formula

Peptone	10 g
Sodium chloride	70 g

Directions   Prepare as described for Medium 54 except adjust pH to 7.6 ± 0.1.

57 PEPTONE SALT (10%) BROTH

Formula

| Peptone | 10 g |
| Sodium chloride | 100 g |

Directions Prepare as described for Medium 54 except adjust pH to
7.6 ± 0.1.

58 PEPTONE SALT NITRATE BROTH

Formula

Peptone	10 g
Sodium chloride	30 g
Potassium nitrate	1 g

Directions Prepare as described for Medium 54.

59 PEPTONE WATER

Formula

| Peptone | 10 g |
| Sodium chloride | 5 g |

Directions Dissolve ingredients in 1 litre of distilled water, adjust
reaction so that the pH after sterilization will be 7.2 ± 0.1, distribute in
10 ml volumes in culture tubes, and autoclave at 121° C for 15 minutes.

Dehydrated Peptone Water is available commercially; reconstitute as
directed on the container.

60 PHENOL-RED CARBOHYDRATE BROTHS

Formula

Peptone or trypticase	10	g
Sodium chloride	5	g
Phenol red	0.025	g
Carbohydrate	5	g

Directions Dissolve all ingredients except carbohydrate in 900 ml of
water, adjust reaction so that the pH after sterilization will be 7 ± 0.1,
dispense in 4.5 ml volumes into culture tubes containing small inverted
Durham fermentation tubes, and sterilize by autoclaving at 121° C for
15 minutes. Cool to room temperature and add 0.5 ml of a filter-sterilized
5% aqueous solution of the desired carbohydrate (pH 7.0) to each tube.

Where gas production is not determined in the test, fermentation tubes can be omitted.

Dehydrated Phenol Red Broth base and dehydrated Phenol Red Broths for most of the carbohydrates commonly used in biochemical identification tests are available commercially; prepare as directed on the container.

61 PHENOL-RED EGG-YOLK POLYMYXIN AGAR

Formula for basal medium

Meat extract	1	g
Peptone	10	g
d-Mannitol	10	g
Sodium chloride	10	g
Phenol red	0.025	g
Agar	15	g

Directions for complete medium   Add ingredients to 1 litre of distilled water, mix well, and heat to boiling to dissolve completely. Cool to 50–60° C and adjust reaction so that the pH after autoclaving will be 7.2 ± 0.1. Dispense in 90 ml volumes into bottles and sterilize at 121° C for 15 minutes. Cool to 45° C and add 10 ml of egg-yolk emulsion (prepared as described in Medium 35 or use a suitable commercial product) and 1 ml of filter-sterilized 0.1% aqueous solution of polymyxin B sulphate.

62 PHOSPHATE-BUFFERED DILUTION WATER (Butterfield, 1932)

(*a*) *Stock phosphate buffer solution*   Dissolve 34 g of potassium dihydrogen phosphate in 500 ml of distilled water. Adjust the pH to 7.2 with 1N NaOH (about 175 ml required) and dilute to 1 litre with distilled water. Sterilize the solution in the autoclave at 121° C for 15 minutes and store in the refrigerator until needed. Final reaction after autoclaving should be pH 7.0 ± 0.1.

(*b*) *Buffered dilution water*   Add 1.25 ml of stock phosphate buffer ((*a*) above) to 1 litre of distilled water. For dilution bottles or tubes (blanks), fill with predetermined volume so that after autoclaving volume will be ±2% of that desired; or, if containers are calibrated, readjust volume aseptically by pipette after autoclaving. Autoclave at 121° C for 15 minutes.

63 RABBIT PLASMA

It is recommended that Difco dehydrated Bacto-coagulase Plasma (Catalogue No. B286) be used. To rehydrate, dissolve the contents of one ampoule (100 mg) in 3.0 ml of sterile distilled water. Unused rehydrated plasma will keep in the refrigerator (0–5° C) for several days.

If dehydrated product is not available, use fresh sterile rabbit or human plasma diluted 1:3 with sterile distilled water. Test each batch with coagulase-positive (include both weak and strong coagulase producers) and coagulase-negative strains of staphylococci before putting it into routine use.

64 REINFORCED CLOSTRIDIAL AGAR (RCM Agar; Hirsch and Grinsted, 1954)

Formula

Yeast extract	3	g
Beef extract	10	g
Peptone	10	g
Glucose	5	g
Soluble starch	1	g
Sodium chloride	5	g
Sodium acetate	3	g
Cysteine hydrochloride	0.5	g
Agar	15	g

Directions Suspend ingredients in 1 litre of distilled water and heat to boiling with frequent agitation until solution is complete. Cool to 50–60° C and adjust reaction so that the pH after autoclaving is approximately 7.0. Sterilize by autoclaving for 20 minutes at 115° C. Cool to about 50° C and pour in 20 ml volumes into Petri dishes equipped with glazed porcelain or Brewer aluminium covers. Dry plates thoroughly before use (see step 9 in introduction to this chapter).

RCM Agar is available commercially in dehydrated form; prepare as described on the container.

65 RINGER SOLUTION

Formula (full strength)

Sodium chloride	9	g
Potassium chloride	0.42	g
Calcium chloride (anhyd.)	0.48	g
Sodium bicarbonate	0.20	g

Directions Dissolve ingredients in 1 litre of distilled water. Dispense as required and sterilize at 121° C for 10 minutes. Final pH, 7.0.

Quarter-strength Dilute one part full strength with three parts distilled water. Mix, dispense as required, and sterilize by autoclaving at 121° C for 10 minutes.

Ringer Solution tablets are available commercially; prepare solution as directed on the container.

66 SALMONELLA-SHIGELLA AGAR

Formula

Beef extract	5.0	g
Peptone or tryptose	5.0	g
Lactose	10.0	g
Bile salts	8.5	g
Sodium citrate	8.5	g
Sodium thiosulphate	8.5	g
Ferric citrate	1.0	g
Agar	13.5	g
Brilliant green	0.00033	g
Neutral red	0.025	g

Directions Suspend ingredients in 1 litre of distilled water, mix well, and heat to boiling with frequent agitation to obtain complete solution. Cool to 45–50° C and pour in 15–20 ml amounts into Petri dishes. Final pH, approximately 7.0. *Do not autoclave.*

The medium is available commercially in dehydrated form; prepare as directed on the container.

67 SALT BROTH 6.5%

Formula

Beef extract	3 g
Peptone	5 g
Sodium chloride	65 g

Directions Dissolve ingredients in 1 litre of distilled water. Adjust reaction for a post-sterilization pH of 7.3 ± 0.2, dispense in 7–10 ml quantities into tubes, and sterilize at 121° C for 15 minutes.

68 SALT BROTH 10%

Formula

Beef extract	3 g
Peptone	5 g
Sodium chloride	100 g

Directions Prepare as directed in Medium 67 above.

69 SALTS CARBOHYDRATE BROTH

Formula for basal medium

Peptone	2	g
Sodium sulphate	2.5	g
Ammonium chloride	5	g
Potassium dihydrogen phosphate	0.5	g
Sodium monohydrogen phosphate	1.5	g
Bromthymol blue (0.2% aqueous solution)	12	ml

Directions for complete medium Dissolve ingredients in 1 litre of synthetic sea-water (see formula below), dispense in 5 ml volumes into culture tubes, and autoclave at 121° C for 15 minutes. Cool and add 0.5 ml of a sterile 10% aqueous solution of the desired carbohydrate (adonitol, arabinose, cellobiose, dulcitol, glucose, inositol, lactose, maltose, mannitol, rhamnose, salicin, starch, sucrose, trehalose, or xylose). Sterilize the carbohydrate solutions either by filtration or by autoclaving at 115° C for 10 minutes.

Synthetic sea-water To prepare, dissolve the following ingredients in 1 litre of distilled water:

Sodium chloride	23.4 g
Potassium chloride	0.66 g
Sodium sulphate	3.91 g
Sodium bicarbonate	0.19 g
Magnesium chloride	4.96 g

70 SALT EGG-YOLK AGAR

Formula for basal medium

Beef extract	3 g
Peptone	5 g
Sodium chloride	100 g
Agar	15 g

Directions for complete medium Add ingredients to 1 litre of distilled water, mix, and heat to boiling with frequent agitation to obtain complete solution. Cool to 50–60° C and adjust reaction so that the pH after autoclaving will be 7.5 ± 0.1. Dispense in 90 ml volumes and sterilize at 121° C for 15 minutes. Before use, melt agar, cool to 50° C, and add 10 ml of sterile egg-yolk suspension (prepared as described in Medium 35) to each 90 ml of medium. Mix well, and pour in 15 ml quantities into Petri dishes. Dry surface of plates before inoculation. (See step 9 of the introduction to this chapter for note on drying plates.)

71 SALT PEPTONE GLUCOSE BROTH

Formula

Proteose peptone	10 g
Glucose	10 g
Sodium chloride	5 g

Directions Dissolve ingredients in 1 litre of distilled water and adjust pH to 7.0–7.2. Distribute in 5 ml volumes in culture tubes and autoclave at 121° C for 15 minutes.

72 SELENITE CYSTINE BROTH

Formula

Tryptone	5 g
Lactose	4 g
Sodium monohydrogen phosphate	10 g
Sodium acid selenite	4 g
l-Cystine solution	1 ml

Directions Dissolve all ingredients except cystine in 1 litre of distilled water and dispense in 10 ml volumes into tubes or in 225 ml volumes into flasks as required. Heat in boiling water bath for 10 minutes. Do not autoclave. Cool and add *l*-cystine solution at the rate of 0.1 ml per 10 ml of medium. Final pH, approximately 7.0. Use medium the same day as prepared.

l-Cystine solution Dissolve 0.1 g of *l*-cystine in 15 ml of 1*N* sodium hydroxide solution and dilute to 100 ml with distilled water. Do not autoclave.

Commercially available dehydrated Selenite Broth can be used. Reconstitute as described on the container and add *l*-cystine as described above.

73 SHEEP-BLOOD AGAR

Directions   Use either of the two basal media described in Medium 30. Prepare complete medium as directed in Medium 30, but use defibrinated sheep blood instead of defibrinated horse blood.

74 SIMMONS CITRATE AGAR (Simmons, 1926)

Formula

Magnesium sulphate heptahydrate	0.2	g
Ammonium dehydrogen phosphate	1	g
Potassium monohydrogen phosphate	1	g
Sodium citrate dihydrate	2	g
Sodium chloride	5	g
Bromthymol blue	0.08	g
Agar	15	g

Directions   Add ingredients to 1 litre of distilled water and heat to boiling to obtain complete solution. Distribute in 10 ml volumes in culture tubes and after sterilization (autoclave at 121° C for 15 minutes) slope to give 1 inch butts. Final pH, 6.8–7.0.

The medium is available commercially in dehydrated form; reconstitute as directed on the container.

75 SIM SALT MEDIUM

Formula

Beef extract	3	g
Peptone	30	g
Sodium thiosulphate	0.05	g
Ferric ammonium citrate	0.05	g
Sodium chloride	30	g
Agar	3	g

Directions   Add ingredients to 1 litre of distilled water, heat to boiling with agitation to obtain complete solution, cool to 50–60° C, adjust pH to 7.4 ± 0.2, dispense in 5–7 ml amounts into culture tubes and autoclave at 121° C for 15 minutes.

Dehydrated SIM Medium is available commercially; reconstitute as described on the container and add sodium chloride to give a final concentration of 3%.

76 SPORULATION BROTH

Formula

Trypticase	20 g
Vitamin-free casamino acids	20 g
Sodium thioglycollate	1 g

Directions   Dissolve ingredients in 1 litre of distilled water, mix well, dispense in 10 ml volumes into 150 × 15 mm screw-capped tubes, and sterilize by autoclaving at 121° C for 15 minutes. Immediately prior to use, add to each tube of medium 1 ml of filter-sterilized aqueous stock solution of thiamine hydrochloride (10 μg per ml). The final concentration of thiamine hydrochloride in each tube is 1.0 μg per ml.

77 STANDARD METHODS AGAR (Plate Count Agar)

Formula

Tryptone	5.0 g
Glucose	1.0 g
Yeast extract	2.5 g
Agar	15 g

Directions   Add ingredients to 1 litre of distilled water, heat to boiling with stirring to obtain complete solution, cool to 45–60° C, and adjust reaction so that the pH after autoclaving will be 7.0 ± 0.1. Dispense as required, and sterilize by autoclaving at 121° C for 15 minutes.

The medium is available commercially in dehydrated form; prepare as directed on the container.

78 STARCH AGAR

Formula

Beef extract	3 g
Peptone	5 g
Soluble starch	2 g
Agar	15 g

Directions   Add all ingredients except starch to 1 litre of distilled water and heat to boiling. Then add starch slowly with vigorous stirring. Boil for 3–4 minutes until solution is complete. Cool to 50–60° C and adjust reaction so that the pH after autoclaving will be 7.2 ± 1. Dispense as required for plates and autoclave at 121° C for 10 minutes.

79 SULPHITE POLYMYXIN SULPHADIAZINE AGAR (SPS Agar)

Formula for basal medium

Tryptone	15	g
Yeast extract	10	g
Iron citrate	0.5	g
Agar	15	g

Directions for complete medium   Add ingredients to 1 litre of distilled water and heat to boiling with agitation to obtain complete solution. Cool to 50–60 ° C, and adjust reaction so that the pH after autoclaving will be 7.0; sterilize at 121° C for 15 minutes. To each litre of sterile medium add aseptically the following filter-sterilized solutions: (1) 5 ml of freshly prepared 10% aqueous solution of sodium sulphite ($Na_2SO_3.7H_2O$), (2) 10 ml of a 0.1% aqueous solution of polymyxin B sulphate, and (3) 10 ml of a 1.2% aqueous solution of sodium sulphadiazine. Mix well and pour into Petri dishes.

80 TELLURITE POLYMYXIN EGG-YOLK AGAR (TPEY Agar; Crisley *et al.*, 1964)

Formula for basal medium

Tryptone	10 g
Yeast extract	5 g
d-Mannitol	5 g
Sodium chloride	20 g
Lithium chloride	2 g
Agar	18 g

Directions for complete medium   Add ingredients to 900 ml of distilled water, heat to boiling with agitation to obtain complete solution, cool to 50–55° C, adjust pH to 7.3, autoclave at 121° C for 15 minutes, and cool to 50–55° C. To this base add aseptically the following ingredients in the amounts stated:

(*a*) 100 ml of egg-yolk emulsion (30% volume/volume egg yolk in physiological saline; to prepare, soak fresh eggs for about 1 minute in 0.1% aqueous solution of mercuric chloride, dry surface with sterile cloth, break open shells, separate the yolks from the whites, and blend yolks mechanically, with desired amount of sterile saline, for 20 seconds, using aseptic precaution throughout).

(*b*) 0.4 ml of a 1% aqueous solution of polymyxin B (sterilize by filtration).

(c) 10 ml of a 1% aqueous solution of potassium tellurite (sterilize by autoclaving at 121° C for 15 minutes).

Mix the medium well and pour in 15–20 ml amounts into sterile Petri dishes.

81 TETRATHIONATE BROTH

Formula for basal medium

Tryptose or proteose peptone	5 g
Bile salts	1 g
Calcium carbonate	10 g
Sodium thiosulphate	30 g

Directions for complete medium   Add all ingredients except iodine solution to 1 litre of distilled water and mix to dissolve soluble materials. Cool and store, if required, at 5–8° C. On the day medium is to be used, add 20 ml of iodine solution (see method of preparation below), agitate gently to mix and to resuspend the precipitate, and dispense aseptically as required. Do not heat this medium after addition of the iodine.

Iodine solution[7]   To prepare, dissolve 5 g of potassium iodide and 6 g of iodine crystals in 10 ml of sterile distilled water in a sterile flask. Dilute to 20 ml with sterile distilled water and store in the dark.

Tetrathionate Broth base is available commercially in dehydrated form; prepare as described on the container.

82 TETRATHIONATE BRILLIANT-GREEN BROTH (Rolfe, 1946)

Formula for basal medium

Beef extract	10 g
Peptone	10 g
Calcium carbonate	50 g

Directions for preparation of basal medium   Add ingredients to 1 litre of distilled water, mix to dissolve soluble materials, adjust pH to 7.4–7.6, and sterilize at 121° C for 15 minutes.

Directions for preparation of complete medium   To each litre of basal medium, add aseptically:

7/Iodine solution is generally added just before the medium is used, but it has been shown that it can be added up to 8 days before use, without significant loss of effect (Galton *et al.*, 1952). This is convenient in field studies, as the completed medium may be dispensed in appropriate containers in the laboratory and transported to the study area ready for use.

(*a*) 1 ml of a sterile 1% aqueous solution of brilliant green dye (sterilize by boiling for 10 minutes).

(*b*) 55 ml of iodine solution (prepare by dissolving 25 g of potassium iodide and 20 g of iodine in 100 ml of distilled water; do not heat). See Medium 81 for note on addition of iodine.

(*c*) 110 ml of a 19.2% aqueous solution of sodium thiosulphate (for convenient measure, dissolve 25 g $Na_2S_2O_3.5H_2O$ in 130 ml of distilled water; sterilize by autoclaving at 121° C for 15 minutes).

Mix well and use from original flask or, if required, dispense aseptically in 10 ml volumes into sterile tubes. Shake flask continually while pouring or pipetting to keep calcium carbonate in suspension.

83 TETRATHIONATE BRILLIANT-GREEN BILE BROTH (Kauffmann, 1935)

Formula for basal medium

Infusion broth	900 ml
Calcium carbonate	50 g

Directions for preparation of complete medium Suspend the calcium carbonate in the infusion broth, adjust pH to 7.4 ± 0.1, autoclave at 115° C for 20 minutes, cool, and add aseptically the following:

(*a*) 100 ml of sterile 50% aqueous solution of sodium thiosulphate ($Na_2S_2O_3.5H_2O$). (Sterilize by autoclaving at 115° C for 25 minutes.)

(*b*) 20 ml of iodine solution (prepare by dissolving 25 g of potassium iodide and 20 g of iodine in 100 ml of distilled water; do not heat). See Medium 81 for note on time of addition of iodine solution.

(*c*) 10 ml of a sterile 0.1% aqueous solution of brilliant green dye (sterilize by boiling for 10 minutes).

(*d*) 50 ml of sterile beef bile (use filtered fresh bile or a 10% aqueous solution of dehydrated ox-gall; sterilize by autoclaving at 115° C for 20 minutes).

Mix well and use from original flask or, if required, dispense aseptically in 10 ml amounts into sterile tubes; shake flask continually while pouring or pipetting to keep calcium carbonate in suspension.

84 TETRATHIONATE BRILLIANT-GREEN BILE SULPHATHIAZOLE BROTH (Galton *et al.*, 1950)

Formula

Complete medium described in Medium 83		
or Medium 81 with added brilliant green (0.01 g)	1000	ml
Sodium sulphathiazole	1.25	mg

Directions   Prepare either of the two basal media as described. Add the sulphathiazole aseptically as a dilute sterile solution to the cooled base medium. For convenient measure, use a 0.0125% aqueous solution of sodium sulphathiazole; add 0.1 ml to 10 ml volumes of base, 1 ml to 100 ml volumes, and 10 ml to 1 litre volumes. Sterilize sodium sulphathiazole solution by boiling for 10 minutes.

85 TETRATHIONATE NOVOBIOCIN BROTH (Jeffries, 1959)

Formula for basal medium

Beef extract	0.9 g
Peptone	4.5 g
Yeast extract	1.8 g
Sodium chloride	4.5 g
Calcium carbonate	25.0 g
Sodium thiosulphate	40.7 g

Directions for complete medium   Add above ingredients to 1 litre of distilled water, mix to dissolve soluble materials, dispense into smaller flasks or tubes as required, and autoclave at 121° C for 15 minutes. Cool medium to below 45° C and add novobiocin at a concentration of 40 μg/ml. The medium is stable for 1 month. Before use, add iodine solution at the rate of 2 ml per 100 ml of medium.

Novobiocin solution   Use sodium salt of novobiocin but calculate weight based on pure novobiocin. For convenient measure, dissolve 0.4 g of novobiocin in 100 ml of distilled water; add 0.1 ml to 10 ml volumes of medium, 1 ml to 100 ml volumes, and 10 ml to 1 litre volumes. Store novobiocin solution in a refrigerator.

Iodine solution   Prepare by dissolving 25 g of potassium iodide and 30 g of iodine in 100 ml of distilled water. See Medium 81 for note on time of addition of iodine solution.

Dehydrated basal medium is available commercially; prepare as directed on the container.

86 THALLOUS ACETATE TETRAZOLIUM GLUCOSE AGAR (Barnes, 1956)

Formula for basal medium

Peptone	10 g
Yeast extract	10 g
Agar	14 g

Directions for complete medium   Add ingredients to 1 litre of distilled water, heat to boiling to obtain complete solution, cool to 50–60° C, adjust reaction so that the pH after autoclaving will be 6.0, distribute in 95 ml volumes in flasks or bottles, and autoclave at 121° C for 20 minutes. Immediately before pouring plates, add aseptically to each 95 ml of melted, tempered (45–50° C) agar base the following solutions in the amounts indicated:

(*a*) 1 ml of a 1% aqueous solution of 2,3,5-triphenyltetrazolium chloride (sterilize by filtration or by steaming for 30 minutes).

(*b*) 5 ml of a 20% aqueous solution of glucose (sterilize by filtration or by steaming for 30 minutes).

(*c*) 2 ml of a 5% aqueous thallous acetate solution (sterilize by autoclaving at 121° C for 15 minutes).

Mix medium well, readjust pH to 6.0 if necessary, and pour in 15–20 ml amounts into sterile Petri dishes.

87 THIOSULPHATE CITRATE BILE SALTS AGAR (TCBS Agar)

Formula

Yeast extract	5 g
Peptone	10 g
Sodium citrate	10 g
Sodium thiosulphate.($5H_2O$)	10 g
Sodium chloride	10 g
Ox-gall (powder)	5 g
Sucrose	10 g
Ferrous citrate	1 g
Sodium cholate	3 g
Agar	15 g
0.2% thymol blue solution	20 ml
0.2% bromthymol blue solution	20 ml

Directions   Add ingredients to 1 litre of distilled water, heat to boiling with agitation to obtain complete solution, cool to 45–50° C, adjust pH to 8.4 ± 0.2, and pour in 15–20 ml volumes into Petri dishes.

Dehydrated TCBS Agar is available from "Nissui" Ltd. and "Eiken" Ltd. (see Medium 5 for complete names and addresses of these companies); reconstitute as directed on the container.

88 TRIPLE SUGAR IRON AGAR

Formula 1

Tryptone or polypeptone	20	g
Sodium chloride	5	g
Lactose	10	g
Sucrose	10	g
Glucose	1	g
Ferrous ammonium sulphate	0.2	g
Sodium thiosulphate	0.2	g
Phenol red	0.025	g
Agar	13	g

Formula 2

Beef extract	3	g
Yeast extract	3	g
Peptone	15	g
Proteose peptone	5	g
Glucose	1	g
Lactose	10	g
Sucrose	10	g
Ferrous sulphate	0.2	g
Sodium chloride	5	g
Sodium thiosulphate	0.3	g
Agar	12	g
Phenol red	0.024	g

Directions Both media are prepared the same way. Suspend ingredients in 1 litre of distilled water, mix well, and heat to boiling with occasional agitation to obtain complete solution. Cool to 50–60° C and adjust pH, if necessary, so that reaction after autoclaving will be of pH 7.3 ± 0.1. Fill 150 × 16 mm tubes one-third full and cap or plug so that aerobic conditions are maintained during use. Autoclave at 121° C for 12 minutes. Cool tubes in slanted position to obtain butts 2.5 cm long and slants 5 cm long (approx.).

Both formulae of Triple Sugar Iron Agar are available commercially in dehydrated form; prepare as directed on the container.

89 TRIPLE SUGAR IRON SALT AGAR

Directions Prepare as described for Triple Sugar Iron Agar (Medium 88, using either formula 1 or 2) but add 30 g of sodium chloride, to give final salt concentration of 3%.

90 TRYPTICASE GLUCOSE MEDIUM

Formula

Trypticase	20	g
Glucose	5	g
Agar	3.5	g
Bromthymol blue	0.01	g

Directions Add ingredients to 1 litre of distilled water, and heat to boiling with frequent agitation to dissolve completely. Cool to 50–60° C and adjust reaction so that the pH after autoclaving will be 7.3 ± 0.1. Dispense into 150 × 15 mm screw-capped tubes filling them half full. Sterilize at 115° C for 15 minutes. Store at room temperature. For growth of clostridia, if the medium is not used the same day it is prepared, heat tubes in flowing steam or in a boiling water bath for a few minutes to drive off dissolved gases and allow to cool without agitation.

Trypticase Glucose Medium is available commercially in dehydrated form; prepare as directed on the container.

91 TRYPTOSE AGAR

Formula

Tryptose	10	g
Glucose	1	g
Sodium chloride	5	g
Agar	15	g
Thiamine hydrochloride	0.005	g

Directions Add ingredients to 1 litre of distilled water, heat to boiling to obtain complete solution, cool to 50–60° C, and adjust reaction so that the pH after sterilization will be 6.9–7.0. Distribute in bottles or tubes as required and sterilize in the autoclave at 121° C for 15 minutes.

Tryptose Agar is available commercially in dehydrated form; reconstitute as directed on the container.

92 TRYPTOSE BILE BROTH 40%

Formula

Tryptose	10	g
Glucose	1	g
Sodium chloride	5	g
Thiamine hydrochloride	0.005	g
Ox-gall	400	g

Directions   Dissolve ingredients in 1 litre of distilled water, and adjust reaction so that the pH after autoclaving will be 6.9–7.0. Dispense in 5 ml volumes into culture tubes and sterilize by autoclaving at 121° C for 15 minutes.

93 TRYPTONE BROTH

Directions   Dissolve 10 g of tryptone in 1 litre of distilled water, dispense in 5 ml portions into culture tubes, and sterilize in an autoclave at 121° C for 15 minutes.

Tryptone Broth is available commercially in dehydrated form; reconstitute as directed on the container.

94 TRYPTOSE BROTH, PH 9.6

Formula

Tryptose	10	g
Glucose	1	g
Sodium chloride	5	g
Thiamine hydrochloride	0.005	g

Directions   Dissolve ingredients in 1 litre of distilled water and adjust reaction (use concentrated sodium hydroxide solution) so that the pH after autoclaving will be 9.6. Dispense in 5 ml volumes into culture tubes and sterilize in the autoclave at 121° C for 15 minutes.

Tryptose Broth is available commercially in dehydrated form; reconstitute as directed on the container and adjust pH before autoclaving.

95 TRYPTOSE SALT BROTH

Formula

Tryptose	10	g
Glucose	1	g
Sodium chloride	65	g
Thiamine hydrochloride	0.005	g

Directions   To prepare, follow directions given for Medium 94 above except adjust pH to 6.9–7.0.

Tryptose Broth is available commercially in dehydrated form; reconstitute as directed on the container and add the extra sodium chloride.

96 TRYPTICASE SOY BROTH WITH 10% SODIUM CHLORIDE

Formula

Trypticase	17	g
Phytone	3	g
Sodium chloride	100	g
Potassium monohydrogen phosphate	2.5	g
Glucose	2.5	g

Directions Dissolve ingredients in 1 litre of distilled water, and dispense into 150 × 15 mm tubes to a depth of 5–8 cm. Autoclave 15 minutes at 121° C. Final pH, 7.2 ± 0.1.

Commercially available dehydrated Trypticase Soy Broth can be used; prepare as directed on the container and add the extra sodium chloride.

97 TRYPTONE SULPHITE NEOMYCIN AGAR (Marshall *et al.*, 1965)

Formula for basal medium

Tryptone	15 g
Yeast extract	10 g
Ferric citrate	5 g
Sodium sulphite	1 g
Agar	15 g

Directions for complete medium Add ingredients to 1 litre of distilled water, heat to boiling with agitation to obtain complete solution, cool to 50–55° C, adjust pH to 7.4, and autoclave at 121° C for 15 minutes. Cool to 47° C and add aseptically 20 ml of a filter-sterilized 0.1% aqueous solution of polymyxin B sulphate (final concentration in medium, 20 μg/ml) and 50 ml of a filter-sterilized 0.1% aqueous solution of neomycin sulphate (final concentration in medium, 50 μg/ml). Prepare on day of use. Use in 23 ml volumes in plastic pouches.

98 TRYPTOSE TELLURITE AGAR

Formula

Tryptose	10	g
Glucose	1	g
Sodium chloride	5	g
Potassium tellurite	5	g
Agar	15	g
Thiamine hydrochloride	0.005	g

Directions   Add ingredients to 1 litre of distilled water, heat to boiling with frequent agitation to dissolve completely, cool to 50–60° C, and adjust reaction for a post-sterilization pH of 6.9–7.0. Distribute as required and autoclave at 121° C for 15 minutes.

Tryptose Agar is available commercially in dehydrated form; reconstitute as directed on the container and add the potassium tellurite.

99 TRYPTOSE TTC AGAR

Formula

Tryptose	10	g
Glucose	1	g
Sodium chloride	5	g
Triphenol tetrazolium chloride	0.1	g
Agar	15	g
Thiamine hydrochloride	0.005	g

Directions   Add ingredients to 1 litre of distilled water, heat to boiling to dissolve completely, cool to 50–60° C, and adjust reaction for a post-sterilization pH of 6.9–7.0. Distribute as required and autoclave at 121ᵛ C for 15 minutes.

Dehydrated Tryptose Agar is available commercially; reconstitute as directed on the container and add the triphenol tetrazolium chloride.

100 VIOLET-RED BILE AGAR

Formula

Yeast extract	3.0	g
Peptone	7.0	g
Sodium chloride	5.0	g
Bile salts No. 3	1.5	g
Lactose	10	g
Neutral red	0.03	g
Crystal violet	0.002	g
Agar	15	g

Directions   Add ingredients to 1 litre of distilled water (the dyes are most conveniently added as 0.1% filtered aqueous solutions), heat to boiling until solution is complete, cool to 50–60° C, adjust reaction so that the pH after autoclaving will be 7.4, distribute as required, and sterilize by autoclaving at 121° C for 15 minutes.

Dehydrated Violet-Red Bile Agar is available commercially; reconstitute as directed on the container.

101 VIOLET-RED BILE GLUCOSE AGAR

Formula

Same ingredients as for Medium 100 above, plus

Glucose	10	g

Directions Add ingredients to 1 litre of distilled water, (add dyes as filtered 1% aqueous solutions) and heat to boiling with agitation until all ingredients are dissolved. Do not sterilize this medium further. Cool to 45° C and pour into Petri dishes, preferably of 15 cm diameter. If not used immediately, store plates in a refrigerator at 5–8° C.

This medium is available commercially in dehydrated form; reconstitute as directed on the container.

102 VOGEL-JOHNSON AGAR (Vogel and Johnson, 1960)

Formula

Tryptone	10	g
Yeast extract	5	g
Mannitol	10	g
Potassium monohydrogen phosphate	5	g
Lithium chloride hexahydrate	5	g
Glycine	10	g
Agar	16	g
Phenol red	0.025	g
Potassium tellurite	0.2	g

Directions Add all ingredients except potassium tellurite to 1 litre of distilled water, mix thoroughly, heat with agitation, and boil 1 minute. Dispense in 100 ml volumes into bottles, autoclave 15 minutes at 121° C, and cool to 45–50° C. To each 100 ml portion, add 2 ml of a 1% potassium tellurite solution which has been autoclaved separately for 15 minutes at 121° C. Mix gently and pour in 15 ml amounts into Petri dishes. Final pH, 7.2 ± 0.2.

Dehydrated Vogel-Johnson Agar is available from Difco Laboratories Inc., Detroit, Mich.; prepare as directed on the container.

103 WATER-BLUE ALIZARINE-YELLOW AGAR (WA AGAR)

Formula

Beef extract	5	g
Peptone	10	g

Sodium chloride	30	g
Sucrose	10	g
Teepol[8]	2	ml
Water blue	0.02	g
Arizaran yellow	0.05	g
Agar	15	g

Directions   Add ingredients to 1 litre of distilled water, heat to boiling with agitation to obtain complete solution, cool to 45–50° C, adjust pH to 7.2 ± 0.1, and pour in 15–20 ml amounts into Petri dishes.

Dehydrated WA Agar is available from "Nissui" Ltd. and "Eiken" Ltd. (see Medium 5 for complete names and addresses of these companies); reconstitute as directed on the container. .

104 XYLOSE LYSINE DESOXYCHOLATE AGAR (Taylor, 1965)

Formula for basal medium

Xylose	3.75	g
l-Lysine HCl	5	g
Lactose	7.5	g
Sucrose	7.5	g
Sodium chloride	5	g
Yeast extract	3	g
Phenol red	0.08	g
Agar	15	g

Directions for complete medium   Add ingredients to 1 litre of distilled water, heat to boiling to obtain complete solution, cool to 50–55° C, adjust reaction so that pH after sterilization will be 6.9, and autoclave at 121° C for 15 minutes. Cool to 50–55° C and add aseptically the following solutions in the amounts indicated:

(*a*) 20 ml of thiosulphate-citrate solution (to prepare, dissolve 34 g of sodium thiosulphate and 4 g of ferric ammonium citrate in 100 ml of water; sterilize by filtration).

(*b*) 25 ml of a 10% aqueous solution of sodium desoxycholate (sterilize by filtration or by autoclaving at 121° C for 15 minutes).

Mix medium well, readjust pH if necessary to 6.9, and pour in 15–20 ml amounts into Petri dishes.

8/Mixture of sodium salts of sulphated fatty alcohols, available from Shell Oil Co.

SPECIFICATIONS FOR
DIAGNOSTIC REAGENTS

1 GRAM'S STAINING METHOD (HUCKER'S MODIFICATION)

There are numerous modifications of the Gram's staining method (Hucker and Conn, 1923, 1927). The Hucker modification is especially valuable for the examination of smears of isolated cultures (Soc. Am. Bact., 1957; APHA, 1966).

CRYSTAL VIOLET SOLUTION

Formula

Crystal violet (85–90% dye content)	2	g
Ethyl alcohol (95%)	20	ml
Ammonium oxalate	0.8	g
Distilled water	80	ml

Directions Dissolve the crystal violet in the alcohol and the ammonium oxalate in the distilled water. Mix the two solutions and store the mixture for 24 hours before use.

IODINE SOLUTION (Burke, 1921)[1]

Formula

Iodine	1	g
Potassium iodide	2	g
Distilled water	100	g

Directions Grind the potassium iodide and iodine together in a mortar adding small increments of water while grinding. Rinse the resulting solution into a volumetric flask and bring the volume to 100 ml.

1/Gram's modification of Lugol's iodine solution can also be used. It is the same as the Burke solution except that 300 ml of water are used instead of 100 ml.

COUNTERSTAIN

Formula

Safranin O	0.25	g
Ethyl alcohol	10	ml
Distilled water	100	ml

Directions Dissolve the safranin in the ethyl alcohol and mix the resultant solution with the distilled water.

STAINING PROCEDURE

(*a*) Stain smear 1 minute with crystal violet.

(*b*) Wash slide gently for a few seconds with water.

(*c*) Flush with iodine solution; allow to stand for 1 minute.

(*d*) Wash gently with water.

(*e*) Allow slide to dry.[2]

(*f*) Wash with successive applications of 95% ethyl alcohol until smear ceases to give off dye (usually three applications are sufficient; total time 30 seconds to 1 minute).

(*g*) Wash with water.

(*h*) Apply counterstain for 10 seconds.

(*i*) Blot dry and examine.

Gram-positive organisms stain blue; Gram-negative organisms stain red.

2 INDOLE PAPER

To strips of filter paper, approximately 0.5 × 5 cm in size, lying flat over blotting paper, add dropwise a solution consisting of 5 g of *p*-dimethyl-aminobenzaldehyde, 50 ml of methanol, and 10 ml of *o*-phosphoric acid. After strips are thoroughly wet, dry them at 70° C. Caution should be taken not to permit these strips to touch the lead acetate paper strips in tubes of Gillies Medium 2.

3 INDOLE REAGENT (Kovacs, 1928)

Formula

Paradimethylaminobenzaldehyde	5	g
Isoamyl (or normal amyl) alcohol	75	ml
Hydrochloric acid (conc.)	25	ml

2/This step is optional. Some laboratories find that better results are obtained by drying before decolorization (Bartholomew and Mittwer, 1952).

Directions   Dissolve the benzaldehyde in the isoamyl alcohol and add the hydrochloric acid. Use 0.2–0.3 ml in each indole test.

Note   Test the paradimethylaminobenzaldehyde for effectiveness, since some brands are not satisfactory and some good brands deteriorate with age. Both amyl alcohol and benzaldehyde should be purchased frequently in amounts consistent with the volume of work to be done.

4 LEAD ACETATE PAPER

To strips of filter paper, approximately 0.5×5 cm in size, lying flat over blotting paper, add dropwise saturated aqueous lead acetate solution until strips are completely wet. Dry the strips at 70° C.

5 LUGOL'S IODINE SOLUTION (Gram's modification)

Formula

Iodine	1 g
Potassium iodide	2 g
Distilled water	300 ml

Directions   Add the chemicals to the distilled water, mix well, and allow 24 hours for the iodine to dissolve. If necessary, add a few more crystals of potassium iodide.

6 METHYL RED SOLUTION

Formula

Methyl red	0.1	g
Ethyl alcohol (95%)	300	ml
Water	200	ml

Directions   Dissolve the methyl red in the alcohol and dilute with the water. For the methyl red test add 5 drops of methyl red solution to 5 ml of the test culture.

7 NAPHTHOL SOLUTION

Formula

α-Naphthol (melting point 92.5° C or higher)	5 g
Absolute ethyl alcohol	100 ml

Directions   Prepare fresh each day. Dissolve the naphthol in the alcohol. In the Voges-Proskauer test, 0.6 ml of naphthol solution and 0.2 ml of a

40% aqueous solution of potassium hydroxide are added to 1 ml of culture.

8 α-NAPHTHYLAMINE SOLUTION

Formula

α-Naphthylamine	0.5 g
5N acetic acid (1 part glacial acetic acid to 2.5 parts of water)	100 ml

Directions Dissolve α-naphthylamine in the acid. In the nitrate test, add equal amounts (0.5–1 ml) of this reagent and Reagent 9 below to each tube.

9 SULPHANILIC ACID SOLUTION

Formula

Sulphanilic acid	1 g
5N acetic acid (1 part of glacial acetic acid to 2.5 parts of water)	125 ml

Directions Dissolve the sulphanilic acid in the acetic acid (dissolves slowly). In the nitrate test, add equal amounts (0.5–1 ml) of this reagent and Reagent 8 above to each tube.

REFERENCES

AMERICAN PUBLIC HEALTH ASSOCIATION. 1960. Standard methods for the examination of dairy products (11th ed.; New York: American Public Health Association, Inc.).

AMERICAN PUBLIC HEALTH ASSOCIATION, AMERICAN WATER WORKS ASSOCIATION, and WATER POLLUTION CONTROL FEDERATION. 1965. Standard methods for the examination of water and wastewater (12th ed.; New York: American Public Health Association, Inc.).

AMERICAN PUBLIC HEALTH ASSOCIATION, SUBCOMMITTEE ON METHODS FOR THE MICROBIOLOGICAL EXAMINATION OF FOODS. 1966. Recommended methods for the microbiological examination of foods (2nd ed.; New York: American Public Health Association, Inc.).

BAIRD-PARKER, A. C. 1962a. An improved diagnostic and selective medium for isolating coagulase-positive staphylococci. J. Appl. Bacteriol. 25, 12.

——— 1962b. The performance of an egg yolk–tellurite medium in practical use. J. Appl. Bacteriol. 25, 441.

BALTIMORE BIOLOGICAL LABORATORY, INC. 1956. BBL products for the microbiological laboratory (4th ed.; Baltimore, Md.: Baltimore Biological Laboratory).

BARNES, E. M. 1956. Methods for the isolation of faecal streptococci (Lancefield Group D) from bacon factories. J. Appl. Bacteriol. 19, 193.

BARTHOLOMEW, J. W. and MITTWER, T. 1952. The Gram stain. Bacteriol. Rev. 16, 1.

BILLING, E. and LUCKHURST, E. R. 1957. A simplified method for the preparation of egg yolk media. J. Appl. Bacteriol. 20, 90.

BURKE, V. 1921. The Gram stain in the diagnosis of chronic gonorrhea. J. Am. Med. Assoc. 77, 1020.

BUTTERFIELD, C. T. 1932. The selection of a dilution water for bacteriological examinations. J. Bacteriol. 23, 355.

CRISLEY, F. D., ANGELOTTI, R., and FOTER, M. J. 1964. Multiplication of Staphylococcus aureus in synthetic cream fillings and pies. Public Health Rept. (U.S.) 79, 369.

DIFCO LABORATORIES INCORPORATED. 1953. The Difco Manual of dehydrated culture media and reagents for microbiological and clinical laboratory procedures (9th ed.; Detroit, Mich.: Difco Laboratories Incorporated).

——— 1962. Difco supplementary literature (Detroit, Mich.: Difco Laboratories Incorporated).

DOLMAN, C. E. 1957. Recent observations on type E botulism. Can. J. Public Health 48, 187.

EDWARDS, P. R. and FIFE, M.A. 1961. Lysine-iron agar in the detection of *Arizona* cultures. Appl. Microbiol. *9*, 478.

GALTON, M. M., STUCKER, C L., MCELRATH, H. B., and HARDY, A. V. 1950. A preliminary study of tetrathionate brilliant green bile broth with added sulfathiazole for the isolation of *Salmonella* from dogs. Bacteriol. Proc. 115.

GALTON, M. M., SCATTERDAY, J. E., and HARDY, A. V. 1952. Salmonellosis in dogs. I. Bacteriological, epidemiological and clinical considerations. J. Infect. Diseases *91*: 1.

GALTON, M. M., LOWERY, W. D., and HARDY, A. V. 1954. *Salmonella* in fresh and smoked pork sausage. J. Infect. Diseases *95*, 232.

GILLIES, R. R. 1956. An evaluation of two composite media for preliminary identification of *Shigella* and *Salmonella*. J. Clin. Pathol. *9*, 368.

HIRSCH, A. and GRINSTED, E. 1954. Methods for the growth and enumeration of anaerobic sporeformers from cheese, with observations on the effect of nisin. J. Dairy Res. *21*, 101.

HUCKER, G. J. and CONN, H. J. 1923. Methods of Gram staining (N.Y. State Agric. Expt. Sta. Tech. Bull. No. 93).

—— 1927. Further studies on the methods of Gram staining (N.Y. State Agric. Expt. Sta. Tech. Bull. 128), p. 3.

HUGH, R. and LEIFSON, E. 1953. The taxonomic significance of fermentative versus oxidative metabolism of carbohydrates by various Gram-negative bacteria. J. Bacteriol. *66*, 24.

HYNES, M. 1942. The isolation of intestinal pathogens by selective media. J. Pathol. Bacteriol. *54*, 193.

JEFFRIES, L. 1959. Novobiocin tetrathionate broth: A medium of improved selectivity for the isolation of salmonellae from faeces. J. Clin. Pathol. *12*, 568.

JORDAN, E. O. and HARMON, P. H. 1928. A new differential medium for the paratyphoid group. J. Infect. Diseases *42*, 238.

KAUFFMANN, F. 1935. Weiter Erfahrungen mit dem kombinierten Anreicherungsverfahren für Salmonellabacillen. Z. Hyg. Infektionskr. *117*, 26.

KOSER, S. A. 1923. Utilization of the salts of organic acids by the colon-aerogenes group. J. Bacteriol. *8*, 493.

KOVACS, N. 1928. A simplified method for detecting indole formation by bacteria. Z. Immunitätsforsch. *56*, 311; Chem. Abs. *22*, 3425.

MARSHALL, R. S., STEENBERGEN, J. F., and MCCLUNG, L. S. 1965. Rapid technique for the enumeration of *Clostridium perfringens*. Appl. Microbiol. *13*, 559.

MCCLUNG, L. S. and TOABE, R. 1947. The egg yolk plate reaction for the presumptive diagnosis of *clostridium sporogenes* and certain species of the gangrene and botulinum groups. J. Bacteriol. *53*, 139.

MCCOY, J. H. 1962. The isolation of salmonellae. J. Appl. Bacteriol. *25*, 213.

NEFEDJEVA, M. P. 1964. Laboratornaia diannostika infektsionnykh zabolevanii [Laboratory diagnosis of infectious diseases; methodological manual] (2nd ed.; Moscow: Biuro Nauchnoi Informatsii), p. 352.

OXOID DIVISION, OXO LTD. 1965. The Oxoid Manual of culture media including ingredients and other laboratory services (3rd ed.; Southward Bridge Rd., London, S.E. 1; Oxoid Division, Oxo Ltd.).

PACKER, R. A. 1943. The use of sodium azide (NaN₃) and crystal violet in a selective medium for streptococci and *Erysipelothrix rhusiopathiae*. J. Bacteriol. *46*, 343.

ROLFE, V. 1946. A note on the preparation of tetrathionate broth. Mon. Bull. Emerg. Public Health Lab. Serv. *5*, 158.

SIMMONS, J. S. 1926. A culture medium for differentiating organisms of the typhoid-colon-aerogenes groups and for the isolation of certain fungi. J. Infect. Diseases *39*, 209.

SMITH, B. A. and BAIRD-PARKER, A. C. 1964. The use of sulphamezathine for inhibiting *Proteus* spp. in Baird-Parker's isolation medium for *Staphylococcus aureus*. J. Appl. Bacteriol. *27*, 78.

SOCIETY OF AMERICAN BACTERIOLOGISTS, COMMITTEE ON BACTERIOLOGICAL TECHNIC. 1957. Manual of microbiological methods (New York: McGraw-Hill).

STRAKA, R. P. and STOKES, J. L. 1957. Rapid destruction of bacteria in commonly used diluents and its elimination. Appl. Microbiol. *5*, 21.

TAYLOR, W. I. 1965. Isolation of shigellae. I. Xylose lysine agars; new media for isolation of enteric pathogens. Am. J. Clin. Pathol. *44*, 471.

VOGEL, R. A. and JOHNSON, M. 1960. A modification of the tellurite glycine medium for use in the identification of *Staphylococcus aureus*. Public Health Lab. *18*, 131.

WILLIS, A. T. and HOBBS, G. 1959. Some new media for the isolation and identification of clostridia. J. Pathol. Bacteriol. *77*, 511.

APPENDICES

APPENDIX I

Agreed Programme of the IAMS International Committee on Microbiological Specifications for Foods

The over-all purpose is to appraise public health aspects of the microbiological content of foods, particularly those of international interest, and to make appropriate recommendations to aid in establishing internationally analytical methods and guides to interpretation of the significance of microbiological data. It is hoped that developing countries will thereby derive particular benefit even though the initial motives for organization of the Committee were based upon existing needs within countries with more advanced food technology.

The Committee has agreed upon the following specific functions:

1 To offer recommendations for limits of tolerance for pertinent categories of microorganisms in specific foods.

2 To recommend and to define methods of analysis and sampling in relation to food-borne microorganisms for which numerical limits are proposed. Categories for which the Committee has already made recommendations include: standard plate count, staphylococci, salmonellae, coliforms.

3 Where sufficient facts are not available, to establish a system of collaborative testing of methods for microbiological examination of foods, with special application to microbiological limits, or to microbial species of immediate public health concern. Of current interest are methods for: salmonellae in diverse foods, enterotoxigenic staphylococci, *Bacillus cereus*, *Clostridium botulinum*, *Clostridium perfringens*, the Enterobacteriaceae, coliforms, faecal coliforms, standard plate count, anaerobic count.

4 To establish channels of organization to aid the interchange of scientists to study specific methods in microbiological analysis.

5 To consider the desirability and feasibility of establishing an international centre or centres for reporting the outbreaks of food-borne illnesses and the determinations from specific foods of food-borne pathogens potentially capable of international spread, such as *Salmonella*, *Vibrio comma*, *Clostridium botulinum*, enteropathogenic viruses, and animal viruses such as rinderpest and hoof and mouth disease.

6 To discuss and to formulate Group Statements upon the significance of specific pathogenic bacteria in foods, such as salmonellae in animal feeds and feed ingredients, staphylococci in dairy products, type E *Clostridium botulinum* in fish and fish products, and viruses in foods.

7 To serve as a consultative body of Food Microbiologists to offer advice, on request, to International Agencies with regard to public health aspects of the microbiological content of foods.

8 To forward the recommendations of the Committee, through its parent body, IAMS, to pertinent International Agencies such as WHO, FAO, UNICEF, as expedient.

9 To maintain liaison with other allied committees such as the committee now engaged in developing the *Codex Alimentarius*.

10 To recommend areas of research for the solution of specific problems in the microbiology of foods, with special reference to the estimation of the significance to public health of (*a*) the presence and source of specific microorganisms and bacterial and fungal toxins in foods, and (*b*) processes which may modify the microbial ecology of foods, such as: (i) the effect of selective antimicrobial agents, such as antibiotics, on microbial populations in foods; (ii) the effect of packaging non-sterile foods in impermeable plastic wrappers; (iii) the significance of the "freeze-dry" process to the microbial content of foods; (iv) the effect of irradiation-pasteurization upon selection and mutagenesis in food-borne microorganisms; (v) the effect of destruction of the normal flora by heat or other agents on subsequent development of pathogenic microorganisms; (vi) the effect of irrigation practices on contamination of raw foods.

11 To establish and maintain liaison between the Committee and sources of Research Grants for microbiological studies within the scope of the Committee.

APPENDIX II

Members of the IAMS International Committee on
Microbiological Specifications for Foods

Dr. F. S. Thatcher (*Chairman*)
Chief, Division of Microbiology, Research Laboratories of Food and Drug Directorate, Department of National Health and Welfare, Ottawa, Ontario, Canada

Dr. D. S. Clark (*Secretary-Treasurer*)
Associate Research Officer, Food Technology Section, Division of Biology, National Research Council, Ottawa, Ontario, Canada

Dr. M. T. Bartram
At the time, Deputy Director, Division of Microbiology, Food and Drug Administration, Department of Health, Education and Welfare, Washington, D.C. 20204, U.S.A. Now retired

Dr. H. E. Bauman
Director of Corporate Research, The Pillsbury Company, 311 Second St. S.E., Minneapolis 14, Minnesota, U.S.A.

Dr. R. Buttiaux
Director, Laboratory of Food Hygiene, Pasteur Institute, 20 Blvd. Louis XIV, Lille (Nord), France

Dr. C. Cominazzini
Director, Department of Microbiology, Provincial Laboratory of Hygiene and Prophylaxis, Via Mossotti 4, 28100 Novara, Italy

Dr. C. E. Dolman
Professor of Bacteriology and Immunology, Department of Microbiology, University of British Columbia, Vancouver 8, B.C., Canada

Mr. R. Paul Elliott
Head, Microbiology Group, U.S. Department of Agriculture, Consumer and Marketing Section, Washington, D.C. 20250, U.S.A.

Mrs. Mildred M. Galton
Chief, Veterinary Public Health Laboratory, Veterinary Public Health Section, Communicable Disease Center, Atlanta, Georgia 30333, U.S.A. (Now deceased)

Dr. H. E. Goresline
Joint FAO/IAEA Division for Atomic Energy in Agriculture, International Atomic Energy Agency, Kaerntnerring, Vienna 1, Austria

Dr. Betty C. Hobbs
Head, Food Hygiene Laboratory, Central Public Health Laboratory, Colindale Avenue, London, N.W. 9, England

Dr. A. Hurst
Unilever Research Laboratory, Sharnbrook, Bedford, England

Dr. H. Iida
Chief, Section of Epidemiology and Assistant Director, Hokkaido Institute of Public Health, South 2, West 15, Sapporo, Hokkaido, Japan

Dr. Keith H. Lewis
Chief, Food Protection, Environmental Sanitation, National Center for Urban and Industrial Health, 222 East Central Parkway, Cincinnati, Ohio 45202, U.S.A.

Dr. H. Lundbeck
Director, The National Bacteriology Laboratory, Box 764, Stockholm, Sweden

Dr. G. Mocquot
Director, Central Station for Dairy Research and Technology of Animal Products, CNRZ, Jouy-en-Josas (S-et-O), France

Dr. D. A. A. Mossel
Head, Laboratory of Bacteriology, Central Institute for Nutrition and Food Research T.N.O., 48 Utrechtseweg, Zeist, The Netherlands

Dr. N. P. Nefedjeva
Laboratory of Microbiology, Institute of Nutrition, Academy of Medical Sciences, Solynka 2/14, Moscow, U.S.S.R.

Dr. Fernando Quevedo
Director del Centro Latinoamericano de Bacteriologia Alimentaria, National University of San Marcos, Puno, Apartado 5653, Lima, Peru

Mr. Bent Simonsen
Chief Microbiologist, Bacteriological Branch of Danish Meat Products Laboratory, Howitzvej 13, Copenhagen F, Denmark

Dr. G. G. Slocum

4204 Dresden Street, Kensington, Maryland 20795, U.S.A. (Formerly Director, Division of Microbiology, Food and Drug Administration, Department of Health, Education and Welfare, Washington, D.C.)

Dr. M. Ingram (ex-officio member, Chairman of Parent Section of Food Microbiology and Hygiene of IAMS)

Director, Meat Research Institute, Langford, near Bristol, Somerset, England

APPENDIX III

Subscribers to the Sustaining Fund of the IAMS International Committee on Microbiological Specifications for Foods
(listed alphabetically)

Beecham Group Ltd., Beecham House, Great West Rd., Brentford, Middlesex, England

Brown and Polson, Ltd., Claygate House, Esher, Surrey, England

Cadbury Schweppes, Ltd., Schweppes House, Grosvenor Rd., St Albans, Herts., England

Carlo Erba Institute for Therapeutic Research, Milan 20159, Italy

Continental Baking Co., P.O. Box 231, Rye, N.Y. 10580, U.S.A.

Distillers Co. Ltd., 21 St. James Square, London, S.W.1, England

Findus Ltd., Bjuv, Sweden

General Foods Corporation, White Plains, N.Y. 10602, U.S.A.

General Foods Ltd., Box 4019, Terminal A, Toronto, Canada

Joseph Rank Ltd., Millcrat House, Eastcheap, London, E.C.3, England

Marks and Spencer Ltd., Michael House, Baker Street, London, W.1, England

Mars Ltd., Dundee Rd. Trading Estate, Slough, SL1 4JX, Bucks., England

McCormick and Co. Inc., Baltimore, Md. 21202, U.S.A.

Meiji Milk Plant, 2–6 Kyobashi, Chuo-ku, Tokyo, Japan

Nihon Eiyokagaku (Eiken) Co., Ltd., 1–33 Hongo, Bunkyo-ku, Tokyo, Japan

Nissui Seiyaku Co., Ltd., 3–22 Sendagi, Bunkyo-ku, Tokyo, Japan

Pasteur Institute of Lille, 20 Louis XIV Blvd., Lille, France

The Pillsbury Co., 311 Second St. S.E., Minneapolis 14, Minn., U.S.A.

Snow Brand Milk Products Co., Ltd., 36 Naebo-cho, Sapporo, Hokkaido, Japan

Sociedad Nacional de Pesqueria, Lima, Peru

Spillers Ltd., Old Change House, 4–6 Cannon Street, London EC4M 6XB, England

Tate and Lyle Refineries Ltd., 21 Mincing Lane, London, England

Terme di Crodo, Via Cristoforo Gluck 35, Milan, 20125, Italy

Tesco Stores Ltd., Tesco House, Delamere Rd., Cheshunt, Waltham Cross, Herts., England

Unilever Ltd., Unilever House, Blackfriars, London, E.C.4, England

World Health Organization, Geneva, Switzerland

Yakult Co., Ltd., 3–6 Nihonbashi-Honcho, Chuo-ku, Tokyo, Japan

APPENDIX IV

Some recommendations on safety precautions
in the microbiological laboratory

Infectious or toxic materials are always potentially dangerous, and should always be treated with due respect and care. Misuse or abuse of such materials in a laboratory can be dangerous, not only to the individual at fault, but to others in his vicinity or even to people widely dispersed where disseminating mechanisms, such as air-conduits, can distribute either pathogens or toxins.

The analysis of foods for pathogens and toxins clearly might involve such contaminative risks, both from the foods and from cultures and toxic concentrates derived from them. Experience establishes that such risks have been minimized almost to the point of extinction by intelligent understanding of the potential hazards and by the application of sound laboratory practice.

The best protection against hazards of the microbiological laboratory is the exercise of common sense in the practice of good laboratory technique. It should be the responsibility of a microbiologist to ensure that his technicians appreciate and use good technique.

The following list of rules comprehends a number of faults known to have caused infection of laboratory personnel. While these rules are of primary importance, they are by no means inclusive. Such rules are of little use unless the laboratory worker uses his common sense as well. For further details on safety procedures for the microbiological laboratory, the reader is referred to the publications listed at the end of this Appendix.

1 Do not eat, drink, or smoke at the laboratory bench.

2 Always wear a laboratory coat.

3 Work benches should have smooth surfaces which can be easily cleaned and disinfected. A splashboard along the wall above the bench is desirable.

4 Wear impermeable gloves when handling infectious or toxic materials; wear goggles when handling botulinum cultures or toxins. Develop the habit of keeping your hands away from your mouth, nose, eyes, and face.

5 Beards are hazards in handling infectious materials. Don't sport one.

6 Subculture infectious organisms in safety cabinets.

7 Cover work areas of the bench with absorbent paper, preferably polyethylene-backed, whenever hazardous material might be spilled.

8 Do not pipette infectious or toxic material by mouth. Plug top ends of pipettes with cotton before sterilization.

9   Immerse used pipettes immediately in a disinfectant. Transfer disinfected pipettes to a container partly filled with soapy water and autoclave; wear impermeable gloves for the transfer.

10   Place all infected material in a proper container for autoclaving. Do not leave the container unattended.

11   Autoclave infected laboratory coats before washing.

12   In case of spills, cover the area immediately with a proper disinfectant. Botulinum toxins should be covered with saturated sodium carbonate.

13   Never moisten labels with the tongue. This may be avoided by using pressure-sensitive labels.

14   After each work phase, wipe the bench thoroughly with a disinfectant, and wash your hands.

15   Centrifuges which are used for centrifugation of toxic or infectious material should be shielded. This is of particular importance for continuous centrifugation.

16   Use only undamaged and capped tubes for centrifugation. Make sure the tubes will not overflow during centrifugation. Avoid decanting centrifuge tubes. If you must decant, wipe the rims with a disinfectant afterwards.

17   Do not use cotton wads when exhausting air and excess liquid from syringe needles. Use instead a small bottle filled with cotton soaked in disinfectant.

18   Before and after injecting animals with infectious material, swab the site of injection with a disinfectant.

19   Use only needle-locking hypodermic syringes.

REFERENCES

CHATIGNY, M. A. 1961. Protection against infection in the microbiological laboratory: Devices and procedures. Advances in Appl. Microbiol. *3*, 131.

MORRIS, E. J. 1960. A survey of safety precautions in the microbiological laboratory. J. Med. Lab. Technol. *17*, 70.

PHILLIPS, G. B. 1961. Microbiological safety in U.S. and foreign laboratories (Technical Study No. 35, Industrial Health and Safety Division, U.S. Army Biological Laboratories, Frederick, Md.).

WEDUM, A. G. 1953. Bacteriological safety. Am. J. Public Health *43*, 1428.

———— 1964. Laboratory safety in research with infectious aerosols. Pub. Health Rept. *79*, 619.

APPENDIX V

Some additional considerations in sampling

ACCURACY OF ATTRIBUTE SAMPLING

Tables exist[1] (for example, Table 10) which shows the probability that $X\%$ or less of a lot is defective if c samples are found to be defective when n random samples are examined from a lot with the total number N. These tables are for use in sampling by attributes only: that is, where one classifies a characteristic as conforming or not-conforming, defective or non-defective (in the present case, with more or less microorganisms than a particular critical level, the "standard" level). The probabilities are based on random sampling. The tables do not apply to restricted samples (unless one is willing to assume that the restricted sample is from a well-"mixed" lot and therefore can be considered to be random).

When the sample is a significant portion of the lot size (as a rule of thumb, more than 10% of the lot size) the probabilities in these tables are technically

TABLE 9

Lot per cent defective $100(p)$	Tabled probability for $n = 29$, $c = 1$	Probability of accepting the lot $(100 -$ tabled probability$)$
1	3	97
2	11	89
3	22	78
4	32	68
5	43	57
6	53	47
7	61	39
8	69	31
9	75	25
10	80	20
12	88	12
14	93	7
16	96	4
18	98	2
20	99	1

1/See *Accuracy of Attribute Sampling. A Guide for Inspection Personnel, 1966*, published by the Consumer and Marketing Service, U.S. Department of Agriculture, Washington, D.C.

TABLE 10

Probabilities for use when the sample size, *n*, is less than 10% of the lot size *N*†

Probability of being right if, based on a sample of *n* which contains *c* defectives, one assumes there is *X*% or less defective in the lot

n	c	1	2	3	4	5	6	7	8	9	10	12	14	16	18	20
											X (%)					
3	0	3	6	9	12	14	17	20	22	25	27	32	36	41	45	49
	1	*	*	*	*	1	1	1	2	2	3	4	5	7	9	10
	2								*					*	1	1
6	0	6	11	17	22	26	31	35	39	43	47	54	60	65	70	74
	1	*	1	1	2	3	5	6	8	10	11	16	20	25	30	34
	2		*	*		*	*	1	1	1	2	3	4	6	8	10
	3													1	1	2
	4															*
13	0	12	23	33	41	49	55	61	66	71	75	81	86	90	92	95
	1	1	3	6	9	14	18	23	28	33	38	47	56	64	71	77
	2	*	*	1	1	2	4	6	8	11	13	20	27	35	42	50
	3			*	*	*	1	1	2	2	3	6	10	14	19	25
	4						*	*	*	*	1	1	3	4	7	10
	5										*	*	1	1	2	3
	6												*	*	*	1
	7															*
21	0	19	35	47	58	66	73	78	83	86	89	93	96	97	98	99
	1	2	7	13	20	28	36	44	51	58	64	74	81	87	91	94
	2	*	1	2	5	8	13	18	23	29	35	47	58	68	76	82
	3	*	*	*	1	2	3	6	8	11	15	24	34	44	54	63
	4		*	*	*	*	1	1	2	4	5	10	16	24	32	41
	5						*	*	1	1	1	3	6	11	16	23
	6								*	*	*	1	2	4	7	11
	7											*	1	1	2	4
	8												*	*	1	1
	9														*	*

n = 29

0	100	100	99	99	98	95	94	91	88	83	77	69	59	44	25
1	99	98	96	93	88	80	75	69	61	53	43	32	22	11	3
2	95	91	86	79	69	57	49	41	33	25	15	11	6	2	*
3	86	79	70	59	47	33	26	20	14	9	5	3	1	*	
4	72	62	50	38	26	16	11	8	5	3	1	1	*		
5	54	43	32	21	13	6	4	3	1	1	*				
6	36	26	17	10	5	2	1	1	*						
7	21	14	8	4	2	1	*								
8	11	6	3	1	1	*									
9	5	3	1	*											
10	2	1	*												
11	1	*													
12	*														

n = 38

0	100	100	100	100	99	98	97	96	94	90	86	79	69	54	32
1	100	100	99	98	95	90	87	82	76	67	57	45	32	18	5
2	99	99	96	92	85	75	68	60	50	40	30	19	10	4	1
3	98	96	88	80	68	54	45	36	27	19	12	6	3	1	*
4	93	90	75	63	49	33	25	18	12	8	4	2	1	*	
5	84	80	58	44	30	17	12	8	5	2	1	*			
6	70	66	41	28	16	8	5	3	1	1	*				
7	54	50	26	15	8	3	2	1	*						
8	37	34	14	8	3	1	1	*							
9	23	22	7	3	1	*									
10	13	12	3	1	*										
11	7	6	1	*											
12	3	3	*												
13	1	1													
14	*														

n = 48

0	100	100	100	100	100	99	99	98	97	95	91	86	77	62	38
1	100	100	99	99	98	96	94	91	86	79	70	58	42	25	8
2	99	100	99	97	94	87	82	75	66	56	43	30	17	7	1
3	98	98	96	92	84	72	64	54	44	33	22	12	6	2	*
4	98	95	90	82	70	53	44	34	24	16	9	4	1	*	
5	94	89	80	68	52	35	26	18	12	6	3	1	*		
6	87	78	67	52	35	20	14	9	5	2	1	*			
7	77	65	51	36	21	10	6	4	2	1	*				

†When the sample size is more than 10% of the lot size, the probabilities in this table will be changed. The extent of change depends on the value of n/N.

not applicable. However, they may still be used as a guide, if one realizes that
the exact probabilities are larger than those given in the tables.

EXAMPLES OF THE USE OF TABLES

Example 1 Assume a random sample of 48 units has been selected from a
large lot and 2 defectives are found in the 48. What is the chance that the lot
contains 5% or less defectives? Referring to Table 10 under $n = 48$, we find the
probability in the row for $c = 2$ and column for $X = 5$ to be 43. Thus there is
a 43% chance that the lot has 5% or less defectives. Faced with this degree
of uncertainty, one would take additional samples if it were important to know
about the 5%. Suppose, however, we only need to know if the lot has 10%

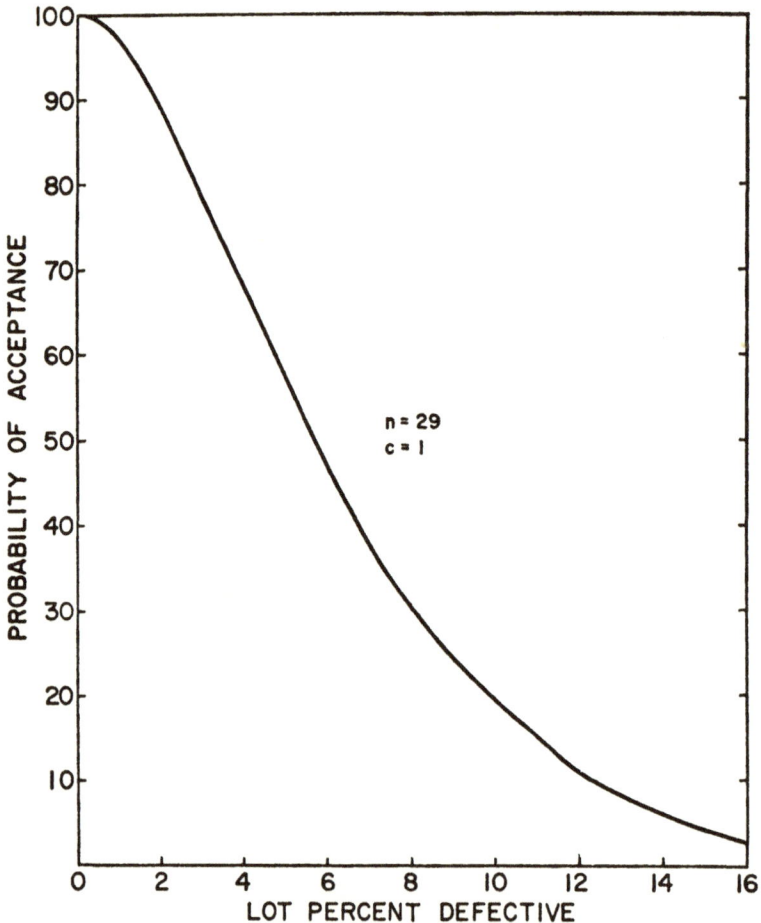

FIGURE 2 Operating characteristic curve for sampling plan $n = 29$, $c = 1$
(with lot size N exceeding 290).

or less defectives. Referring again to the tables under $n = 48$ and opposite $c = 2$, we find the probability of 87 in the $X = 10$ column. Thus we can rely on our sample in concluding the lot has 10% or less defectives, with an 87% chance of being right.

Example 2   Basically the plan is to draw a random sample of 29 units and to accept the lot if none of these or only one is defective. An O.C. (operating characteristic) curve for this plan can be constructed using a sample size (n) of 29 and an acceptance number (c) of 1 (with the lot size, N, greater than 10 times the sample size); the O.C. curve gives the probability of accepting a lot containing a proportion of "p" defectives. The probability of accepting a lot when $n = 29$, $c = 1$, and $X = 100(p)$ can be determined by subtracting the tabled probability from 100. This method gives Table 9 and Fig. 2.

INDEX

Advisory Committee on the Microbiology of Frozen Foods 54
Aerobic mesophilic bacteria, *see* Mesophilic bacteria
Aeromonas (*Oceanomonas*) *parahaemolyticus* 15
Agar, specifications for 151
Agar plate colony count for mesophiles
methods of determination 64–69; Pour Plate method 66–67; Surface Plate method 67–68; Drop Plate method 68–69
value and limitations of 23–25, 64
Antigens: of *C. perfringens* 127; of *V. parahaemolyticus* 108
Antisera
for *C. botulinum* 19, 124
for *C. perfringens* 129, 131
for salmonellae 101–2; polyvalent O antisera 101, 102; individual O antisera 101, 104; polyvalent H antisera 101, 103; Spicer-Edwards H antisera 101, 102, 104, 105
for staphylococci 16
Antitoxins: for *C. botulinum* 124; for *C. perfringens* 129, 130
Arizona 5, 26, 100, 107
Autoclaving of media 157

Bacillus anthracis 3
Bacillus cereus 4, 57; enumeration and identification of 138–41; significance of in foods 22, 138
Baird-Parker Agar: preparation of 158; use of 116–17; appearance of staphylococci on 116–17
Baltimore Biological Laboratories, Inc. 152, 153, 154, 158
Bartholomew and Mittwer's spore stain method 135
Beef Extract, specifications for 151
Bile Salts, specifications for 152

Bile Salts No. 3, specifications for 152
Bismuth Sulphite Agar (McCoy): preparation of 160; use of 92, 93; appearance of salmonellae on 95
Bismuth Sulphite Agar (Wilson and Blair, modified): preparation of 159; use of 92, 93; appearance of salmonellae on 95
Bismuth Sulphite Phenol-Red Agar: preparation of 160; use of 109; appearance of *V. parahaemolyticus* on 109
Bismuth Sulphite Salt Broth: preparation of 161; use of 108, 109
Blendors: specifications for 61, 63; operation of 62, 63
Blood Agar: preparation of 162; use of 119, 120, 126
Botulism 18–20; cause of 18, 19; control of 19–20
Brain Heart Infusion Agar: preparation of 162; use of 126
Brain Heart Infusion Broth: preparation of 163; use of 122
Brilliant Green Agar: preparation of 163; use of 92, 93, 94; appearance of salmonellae on 94
Brilliant-Green Lactose Bile Broth 2%: preparation of 163; use of 72, 73, 76, 78
Brilliant-Green MacConkey Agar: preparation of 164; use of 92, 93; appearance of salmonellae on 94
Brilliant-Green Sulphadiazine Agar: preparation of 164; use of 92, 93, 94, 96; appearance of salmonellae on 94
British Drug House, Ltd. 159
Brucella 3
BTP Teepol Agar: preparation of 165; use of 108, 109; appearance of *V. parahaemolyticus* on 109

www.ingramcontent.com/pod-product-compliance
Lightning Source LLC
Chambersburg PA
CBHW030502210326
41597CB00013B/756